QE 471.15 .S5024 1990
O'Brien, Neal R
Argillaceous rock atlas

DATE DUE

Alpine Campus
Learning Resource
Center
P.O. Box 774688
Steamboat Springs,
CO 80477

DEMCO

Argillaceous Rock Atlas

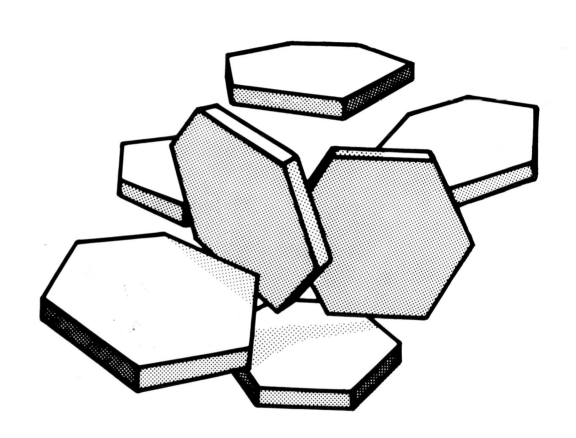

Neal R. O'Brien   Roger M. Slatt

Argillaceous Rock Atlas

With 242 Illustrations, 46 in Full Color

Springer-Verlag
New York Berlin Heidelberg
London Paris Tokyo Hong Kong

NEAL R. O'BRIEN
Geology Department
Potsdam College
State University of New York
Potsdam, NY 13676-2294, USA

ROGER M. SLATT
ARCO Oil and Gas Company
Research and Technical Services
2300 West Plano Parkway
Plano, TX 75075, USA

On the front cover: Bituminous shale formation with pyrite framboid (Upper Lias, Jurassic), p. 32.

Frontispiece: Representation of clay flakes.

Library of Congress Cataloging-in-Publication Data
O'Brien, N. R. (Neal R.)
　　Argillaceous rock atlas / N.R. O'Brien, R.M. Slatt.
　　　　p.　　cm.
　　Includes bibliographical references (p.　　).
　　1. Shale—Atlases.　I. Slatt, Roger M.　II. Title.
QE471.15.S5024　　　1990　　　　　　　　　　　　　　　90-9739
552'.5—dc20　　　　　　　　　　　　　　　　　　　　　　CIP

Printed on acid-free paper

© 1990 Springer-Verlag New York Inc.
All rights reserved. This work may not be translated or copied in whole or in part without the written permission of the publisher (Springer-Verlag New York, Inc., 175 Fifth Avenue, New York, NY 10010, USA), except for brief excerpts in connection with reviews or scholarly analysis. Use in connection with any form of information storage and retrieval, electronic adaptation, computer software, or by similar or dissimilar methodology now known or hereafter developed is forbidden.
The use of general descriptive names, trade names, trademarks, etc., in this publication, even if the former are not especially identified, is not to be taken as a sign that such names, as understood by the Trade Marks and Merchandise Marks Act, may accordingly be used freely by anyone.

Camera-ready copy provided by the authors.
Printed and bound by Arcata Graphics/Kingsport, Kingsport, TN, USA.
Printed in the United States of America.

9 8 7 6 5 4 3 2 1

ISBN 0-387-97306-0 Springer-Verlag New York Berlin Heidelberg
ISBN 3-540-97306-0 Springer-Verlag Berlin Heidelberg New York

Neal dedicates this Atlas to his father, Raymond, and wife, Kathy. Roger dedicates the Atlas to his parents Earl and Helen, for supporting his early interest and education in geology, and to his wife, Linda, and sons, Thomas and Andrew, for their love, patience, and understanding during the long course of this research.

PREFACE

In 1980, the book Sedimentology of Shale, Study Guide and Reference Source (Potter et al., 1980) was published to provide a source book for information and ideas on a class of rocks that was very poorly studied and understood. The authors noted that even though shales comprise 60% of the stratigraphic column, they are often taken for granted and considered the "interbedded matrix" between lithologies of greater economic significance or interest. The authors also noted the importance of shales as sources of oil, gas, and heavy metals in addition to their holding the key to much of earth's history. They acknowledged that the book was unusual in that, rather than summarizing and evaluating the results of years of study of the subject, it was largely devoted to identifying problems in the study of shale and emphasizing areas for expanded research.

A major reason for the lack of understanding of argillaceous rocks is difficulty in analysis. Non-routine methods are required to identify small textural and structural features that can normally be observed with the naked eye or hand lense in coarser-grained rocks. In our book, we have attempted to overcome this obstacle by demonstrating an approach to the systematic analysis of argillaceous rocks. We show that this approach provides a framework for interpreting depositional and post-depositional geological processes and environments. As such, it goes an important step beyond the 1980 source book (Potter et al., 1980) and fills a gap in our understanding of argillaceous rocks.

The chapters are arranged as follows. First we have listed important terms and classification schemes. Then we explain a systematic analytical approach which combines x-radiography, thin-section petrography, and scanning electron microscopy. Much of the remainder of the book is devoted to case studies which illustrate the utility of the techniques and features identified in determining sedimentary processes and environments of deposition. We also present a chapter on the physical conversion of mud to shale. A chapter on fabrics of hydrocarbon source rocks and oil shales and the possible role of fabrics on primary hydrocarbon migration is also presented. Fabric information is supplemented by mineralogical and geochemical data throughout the book, and is also summarized in a single chapter. Finally, we provide some conclusions, point out new questions that have arisen in the course of our studies, and list ideas for future research.

The goal of this book is to provide a visual frame of reference of fabrics to aid geologists in the interpretation and understanding of argillaceous rocks, and to demonstrate that considerable information on sedimentary processes and environments can be interpreted through their systematic and careful analysis.

We wish to thank the many people and organizations who made this book possible. R. W. Dalrymple and J. J. Renton assisted in preparation of the x-radiographs. SEM facilities were provided by E. Matijevic of Clarkson University. Selected samples were obtained from C. W. Atkinson, M. R. Rosen, M. H. Scheihing, J. T. Senftle, J. D. Reed, S. S. Demecs, F. E. Cole, R. M. Coveney, P. J. DeMaris, J. D. Gray, P. H. Heckel, J. W. Hosterman, M. R. House, S. D. Hovorka, Illinois State Geological Survey, Indiana Geological Survey - Department of Natural Resources, Iowa Geological Survey, W. T. Kirchgasser, Ohio Geological Survey, D. Patchen, D. W. Reif, Salt Repository Project Office - U. S. Department of Energy, N. R. Shaffer, R. J. Spencer, J. P. M. Syvitski, E. G. Wermund, ARCO Research & Technical Services, and the many undergraduate members of the Potsdam College Shale Research Team. M. R. House, R. Young, and W. T. Kirchgasser assisted in field work and R. Peppers, A. Traverse, D. W. Powers provided valuable advice about sample identification. J. T. Senftle provided valuable assistance in interpreting geochemical data and in reviewing the chapter on hydrocarbon source rocks and oil shales. Neal R. O'Brien is grateful for financial assistance from the Donors of the Petroleum Research Fund, administered by the American Chemical Society, the National Science Foundation (grant EAR-8611608), and the Potsdam College Mini-grant program. Roger M. Slatt thanks ARCO Oil and Gas Co. for permission to publish this book. K. L. Bianchi, J. M. Richardson and V. Marks expertly and patiently typed several manuscript drafts and L. G. Slatt compiled and organized the photographs.

CONTENTS

PREFACE   vii

CHAPTER 1   INTRODUCTION   1

    Definitions 2

    Argillaceous Rock Classifications 3

    Stratification and Parting Description 4

    Examples Seen in Outcrop 4

    Classification of Very Fine-Grained Sedimentary Rocks 5

    Textural Classification of Fine-Grained Sediments and Rocks 6

CHAPTER 2   FABRIC ANALYSIS TECHNIQUES   7

    X-radiography 8

    Petrography 8

    Scanning Electron Microscopy 9

CHAPTER 3   X-RADIOGRAPHY, PETROGRAPHY AND SCANNING ELECTRON MICROSCOPY DESCRIPTIONS   11

       X-radiography Classification of Argillaceous Rock Macrofabric   12

            Well developed lamination   12

            Indistinct lamination   12

            Bioturbation   12

            Suggested environmental significance of macrofabric x-radiography data   14

       Petrographic Classification of Black Shales   16

            Black shale - lamination types   16

                 Finely laminated   16

                 Thickly laminated   16

                 Wavy laminated   16

                 Lenticular laminated   16

            Various petrographic features of shale - bioturbation   22

            Petrographic features of shale - miscellaneous   23

       Scanning Electron Microscopy Descriptions   24

            Preferred particle orientation in shale   24

            Random particle orientation in mudstone   26

            Fabric variations in organic rich shales   28

                 Microfabric types of organic rich shales   28

CHAPTER 4   MISCELLANEOUS FEATURES IN ARGILLACEOUS ROCKS   31

    Pyrite Framboids   32

    Fecal Pellets   34

    Palynomorphs in Shales   36

CHAPTER 5   CASE STUDIES OF SPECIFIC DISTINCTIVE FEATURES   39

    Well Developed Lamination in a Black Shale (Example I)   40

        Bituminous Shale Formation (Upper Lias, Jurassic)   40

    Well Developed Lamination in a Black Shale (Example II)   42

        Jet Rock Shale Formation (Upper Lias, Jurassic)   42

    Organic Variation in a Shale - Clues to the Cause of Lamination   44

        Jet Rock Shale Formation (Upper Lias, Jurassic)   44

    Laminated Shale from Bottom-Flowing Low Density Turbidity Currents   46

        Sunbury Shale Formation (Mississippian)   46

    Bioturbation   48

        Huron Shale Member (Ohio Shale Formation, Devonian)   48

    Bioturbation - Tiered Burrowing in Shale   50

        Gray Shale Formation (Upper Lias, Jurassic)   50

> Significance of Vertical Fabric Variation in a Shale 52
>> Huron Shale Member (Ohio Shale Formation, Devonian) 52
> A Journey to "Anoxia" - Reconstruction of an Event on the Devonian Sea Floor 54
>> Huron Shale Member (Ohio Shale Formation, Devonian) 54

CHAPTER 6   CASE STUDIES OF FABRIC ANALYSIS IN EVALUATING SEDIMENTARY PROCESSES AND ENVIRONMENTS   57

> Marine Regressional Facies 58
>> Hushpuckney Shale Member (Swope Formation, Pennsylvanian) 58
> Marine Transgressional Facies 60
>> Rhinestreet Member (West Falls Formation) and Cashaqua Member (Sonyea Formation) (Devonian) 60
> Floodplain-Paleosol Facies 62
>> Ivishak Sandstone (Triassic) 62
> Evaporite Facies 64
>> Great Salt Lake (Pleistocene-Holocene), Bristol Dry Lake (Pliocene-Holocene), Clear Fork Formation (Permian) 64
> Tidal Flat Facies 66
>> Red Bed Members (Moenkopi Formation, Lower Triassic) 66

Shallow Marine Shelf Facies 72
- Wilcox Group (Lower Eocene) 72
- Kuparuk River Formation (Lower Cretaceous) 76

Delta Complex Facies 78
- Argillaceous Units of the Ferron Sandstone Member (Mancos Formation, Cretaceous) 78

Submarine Slope Facies 82
- Mudstone Facies, Cozy Dell Formation (Middle Eocene) 82

Marine Turbidite Facies 84
- Huron Shale Member (Ohio Shale Formation, Devonian) 84

Deep Marine Turbidite Facies 86
- Pico Formation (Early Pliocene) 86

Marine Basinal Facies 88
- Geneseo Shale Member (Genesee Formation, Devonian) 88

CHAPTER 7   FORMATION OF SHALE BY COMPACTION OF FLOCCULATED CLAY--A MODEL   91

CHAPTER 8   FABRICS OF SOME HYDROCARBON SOURCE ROCKS AND OIL SHALES   97

    Introduction   98

    Marine Hydrocarbon Source Rock   100

        Kimmeridge Clay (Jurassic)   100

        Woodford Formation (Devonian-Mississippian)   104

        Monterey Formation (Phosphatic Facies, Miocene)   106

    Saline Lacustrine Hydrocarbon Source Rock   110

        Green River Formation (Eocene)   110

    Fresh-Brackish Lacustrine Hydrocarbon Source Rock   112

        Rundle Oil Shale (Eocene-Oligocene)   112

CHAPTER 9   FABRIC OF GEOPRESSURED SHALE   115

    Geopressured Shale Analysis   116

    General Geology and Composition   116

        Onshore Louisiana   116

        High Island, Texas Offshore Area   117

        South Padre Island, Texas Offshore Area   117

    Description of Shale Fabric   117

    Interpretation of Shale Fabrics   117

CHAPTER 10　COMPOSITION OF ARGILLACEOUS ROCKS　121

CHAPTER 11　CONCLUSIONS　129

REFERENCES　133

INDEX　139

CHAPTER 1
INTRODUCTION

DEFINITIONS

It is appropriate at the beginning to define our terms since there is some disagreement or ambiguity concerning precise terminology applied to argillaceous rocks. In this book the following definitions from Bates and Jackson (1987) <u>Glossary of Geology</u> are used:

ARGILLACEOUS "Pertaining to, largely composed of, or containing clay-size particles or clay minerals, such as an "argillaceous ore" in which the gangue is mainly clay; esp., said of a sediment (such as marl) or a sedimentary rock (such as shale) containing an appreciable amount of clay ...."

CLAY "A rock or mineral fragment or a detrital particle of any composition (often a crystalline fragment of a clay mineral), smaller than a very fine silt grain, having a diameter less than 1/256 mm (4 microns or 0.00016 in., or 8 phi units). This size is approximately the upper limit of size of particle that can show colloidal properties . . composed primarily of clay-size or colloidal particles and characterized by high plasticity and by a considerable content of clay minerals and subordinate amounts of finely divided quartz, decomposed feldspar, carbonates, ferruginous matter, and other impurities ...."

FABRIC "[SED] The orientation (or lack of it) in space of the elements (discrete particle, crystals, cement) of which a sedimentary rock is composed."

FISSILITY "A general term for the property possessed by some rocks of splitting easily into thin layers along closely spaced, roughly planar, and approximately parallel surfaces, such as bedding planes in shale or cleavage planes in schist; its presence distinguishes shale from mudstone...."

LAMINA "[SED] The thinnest recognizable unit layer of original deposition in a sediment or sedimentary rock, differing from other layers in color, composition, or particle size; specif. such a sedimentary layer less than 1 cm in thickness (commonly 0.05 - 1.00 mm thick)...."

MUD "[SED] An unconsolidated sediment consisting of clay and/or silt, together with material of other dimensions (such as sand) mixed with water...."

MUDROCK "A <u>syn</u>. of mudstone."

MUDSTONE "An indurated mud having the texture and composition of shale, but lacking its fine lamination or fissility; a blocky or massive, fine-grained sedimentary rock in which the proportions of clay and silt are approximately equal; a nonfissile mud shale."

SHALE "A fine-grained detrital sedimentary rock, formed by the consolidation (esp. by compression) of clay, silt, or mud. It is characterized by finely laminated structure, which imparts a fissility approximately parallel to bedding...."

SILT "[SED] A rock fragment or detrital particle smaller than a very fine sand grain and larger than coarse clay, having a diameter in the range of 1/256 mm to 1/16 mm (4-62 microns, or 0.00016 - 0.0025 in., or 8 to 4 phi units . . .) It varies considerably in composition but commonly has a high content of clay minerals...."

TEXTURE "[PETROLOGY] The general physical appearance or character of a rock, including the geometric aspect of, and the mutual relations among, its component particles or crystals; e.g. the size, shape, and arrangement of the constituent elements of a sedimentary rock...."

ARGILLACEOUS ROCK CLASSIFICATIONS

Following are examples of the terminology used in various argillaceous rock classifications. Notice that there is no agreement upon the use of terms - this is the current state of classifications. In this book, we use the term *mudstone* or *mudrock* when referring to a non-fissile or non-laminated rock and the term *shale* for the argillaceous rock exhibiting fissility or lamination. Potter et al. (1980) presented a classification of partings for what they term *mudrocks* (Note: Their usage of this term differs from ours). This classification and the appearance of these partings are also included below.

STRATIFICATION AND PARTING DESCRIPTION

THICKNESS	STRATIFICATION		PARTING	COMPOSITION
30cm —	THIN	BEDDING	SLABBY	CLAYSTONE AND ORGANIC CONTENT / SAND, SILT AND CARBONATE CONTENT
3cm —	VERY THIN			
10mm —	THICK	LAMINATION	FLAGGY	
5mm —	MEDIUM		PLATY	
1mm —	THIN		FISSILE	
0.5mm —	VERY THIN		PAPERY	

*After Potter et al., 1980

EXAMPLES SEEN IN OUTCROP

CLASSIFICATION OF VERY FINE-GRAINED SEDIMENTARY ROCKS

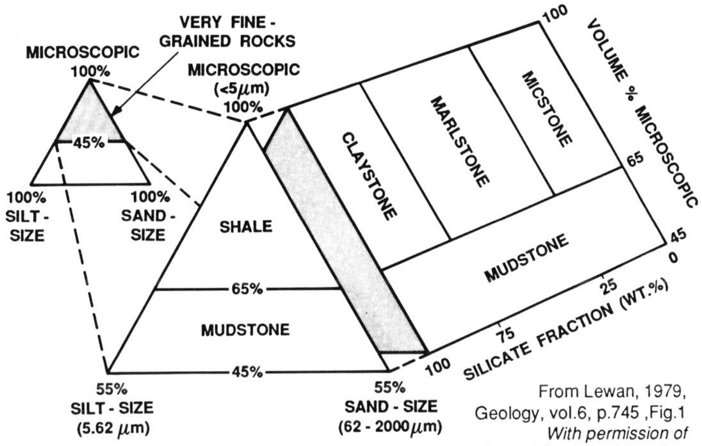

From Lewan, 1979, Geology, vol.6, p.745 ,Fig.1 *With permission of Geological Society of America.*

CLASSIFICATION OF MUDROCKS			
IDEAL SIZE DEFINITION	FIELD CRITERIA	FISSILE MUDROCK	NONFISSILE MUDROCK
> 2/3 SILT	ABUNDANT SILT VISIBLE WITH HAND LENS	SILT - SHALE	SILTSTONE
> 1/3 < 2/3 SILT	FEELS GRITTY WHEN CHEWED	MUD - SHALE	MUDSTONE
> 2/3 CLAY	FEELS SMOOTH WHEN CHEWED	CLAY - SHALE	CLAYSTONE

From Blatt, Middleton & Murray, Origin of Sedimentary Rocks,2e,©1980,p.382. *Adapted by permission of Prentice-Hall,Inc., Englewood Cliffs, New Jersey.*

		SILT FRACTION	2/3	1/3
INDURATED	NON - LAMINATED	SILTSTONE	MUDSTONE	CLAY STONE
	LAMINATED		MUDSHALE	CLAY SHALE

From Lundegard & Samuels,1980 Journal of Sedimentary Petrology, vol.50,p.783,Fig. 1 *With permission of SEPM (Society for Sedimentary Geology).*

PERCENTAGE CLAY - SIZE CONSTITUENTS			0 - 32	33 - 65	66 - 100
FIELD ADJECTIVE			GRITTY	LOAMY	FAT OR SLICK
INDURATED	BEDS	GREATER THAN 10mm	BEDDED SILTSTONE	MUDSTONE	CLAYSTONE
	LAMINAE	LESS THAN 10mm	LAMINATED SILTSTONE	MUDSHALE	CLAYSHALE

Modified after Potter et al. (1980)

TEXTURAL CLASSIFICATION OF FINE-GRAINED SEDIMENTS AND ROCKS

SEDIMENTARY ROCK CONTAINING MORE THAN 50% SILT AND/OR CLAY			
	NO CONNOTATIONS AS TO BREAKING CHARACTERISTICS	MASSIVE	FISSILE
NO CONNOTATIONS AS TO RELATIVE AMOUNTS OF SILT & CLAY	MUDROCK	MUDSTONE	MUDSHALE
SILT > CLAY	SILTROCK	SILTSTONE	SILTSHALE
CLAY > SILT	CLAYROCK	CLAYSTONE	CLAYSHALE

From Ingram, 1953,
Geological Society of America Bulletin, vol.64, p.870, table 1
With permission of Geological Society of America.

CLASSIFICATION OF FINE-GRAINED FRAGMENTAL ROCKS & SEDIMENTS			
	UNCONSOLIDATED AGGREGATE	CONSOLIDATED ROCK	
		GENERAL TERM	NONFISSILE	FISSILE
GENERAL TERM	MUD IF WET, DUST IF DRY	MUDROCK OR LUTITE	MUDSTONE	SHALE OR MUDSHALE
PARTICLES MAINLY >4 MICRONS	SILT	SILTROCK	SILTSTONE	SILT SHALE
PARTICLES MAINLY <4 MICRONS, NORMALLY COMPOSED OF CLAY MINERALS	CLAY	CLAYROCK	CLAYSTONE	CLAY SHALE
VERY WEAKLY METAMORPHOSED	—	—	ARGILLITE	CLAY SLATE

From Dunbar & Rogers, 1957,
Principals of Stratigraphy, p.166, table 10
With permission of John Wiley & Sons, Inc.

MUDROCK DIVISION BASED UPON TEXTURE AND STRUCTURE			
GRAIN SIZE OF MUD FRACTION	SOFT	INDURATED, NONFISSILE	INDURATED, FISSILE
OVER 2/3 SILT	SILT	SILTSTONE	SILT-SHALE
SUBEQUAL SILT AND CLAY	MUD	MUDSTONE	MUD-SHALE
OVER 2/3 CLAY	CLAY	CLAYSTONE	CLAY-SHALE

Used with permission from Folk, 1965,
Petrology of Sedimentary Rocks, p.130.

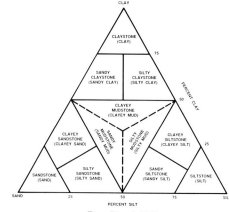

From Picard, 1971
Journal of Sedimentary Petrology, vol.41, p.185, Fig.3
Used with permission of SEPM (Society for Sedimentary Geology).

CHAPTER 2
FABRIC ANALYSIS TECHNIQUES

Based upon our experience, we suggest studying the fabrics of argillaceous rocks by combining three techniques: x-radiography, petrography, and scanning electron microscopy. Each provides useful information of macrofabrics revealed by radiography and thin-sections, and microfabrics revealed with the scanning electron microscope (SEM). Thus, argillaceous rock fabric analysis is a three-step procedure.

Proper sample preparation is essential in any investigation. Listed below are suggestions for obtaining good samples for study. These techniques have been developed by experimentation in the authors' laboratories and are based upon standard preparation techniques combined with suggestions provided by colleagues.

X-RADIOGRAPHY

X-radiography reveals internal fabric of rocks not easily seen by the unaided eye or under a light microscope. Anyone who has had their teeth x-rayed knows the value of this technique in revealing internal structures. In argillaceous rock analysis, slabs of the sample are prepared to a thickness of 2-3mm. Thicker samples may be used, however, this requires varying the radiation time and intensity. The other dimensions of the slab are not critical. The sample should be slabbed perpendicular to bedding or lamination (all of the x-radiographs in this book are oriented in this manner). Some argillaceous rocks are difficult to slab to the correct thickness in a water lubricated diamond saw. Some shales split along fissile planes and certain mudstones disintegrate, often into a clay slurry. This problem can be overcome by using a diamond saw with a kerosene-oil lubricant or simply by dry sawing (although this is a dusty process). To prevent the rock from fragmenting, it is advisable to coat it with a thin layer of standard epoxy resin which dries into a waterproofing layer and holds the rock together after cutting. Often simply wrapping the rock tightly with plastic tape before cutting may prevent it from breaking apart during sawing.

After a 2-3mm slab has been prepared, saw marks must be removed using either fine sandpaper or carborundum powder on a water wet glass plate or grinding wheel. Salad or cooking oil may be substituted for water as a lubricant if the sample breaks apart during wet grinding. Finish the grinding with a 600 grade carborundum powder and observe the sample under a light microscope for the presence of scratches, since they could appear on the final x-radiograph. The finished sample should be of uniform thickness, since a wedge-shaped slab produces an uneven x-radiograph. We usually finish grinding with a 5μm aluminum oxide powder and polish with a polishing compound. The rock slab is then placed on top of x-radiography film and exposed. Instrument settings, distance from source to film, and exposure time must be determined experimentally for the particular instrument. The photos in this book were obtained with an x-radiography unit using a setting of 40 KV for approximately 5 minutes and a source to specimen distance of 60 cm. Positive prints have been made from the x-radiograph negatives. All x-radiographs shown here are positive contact prints or positive enlargements.

More detailed information on x-radiography plus excellent photos may be found in the publications of Bouma (1960), Calvert and Veevers (1962), Clifton (1966), Cluff (1980), Hamblin (1962), and Nuhfer et al. (1979).

PETROGRAPHY

There are some publications concerning the petrography of shales and other argillaceous rocks. Photomicrographs are found in the work of Cole and Picard (1975), Thiessen (1925), and Folk (1962) and provide the reader with useful interpretations. Good color photomicrographs are found in Potter et al. (1980). Anyone who has tried to make a shale or mudstone thin-section can appreciate why so few photomicrographs are in the literature. These thin-sections are simply difficult to make. We have developed a few techniques in addition to the standard procedures which provide useable thin sections. Using a diamond saw, the sample is cut to obtain a surface perpendicular to bedding. Only small rock pieces are needed to make a thin-section (a 4x4x2 cm sample is adequate). If you find the sample remains intact after wet sawing, this is fortunate and allows you to proceed to the next step. Usually, a sample will split upon wet sawing. To prevent this, the sample should be coated with an epoxy resin, then cut with the wet diamond saw.

As mentioned in describing x-radiography techniques, mudstones often break easily, so these samples are usually shaped by dry cutting either with a diamond saw or a simple hack saw (the latter is very tedious and requires patience - but it works).

After cutting, the rock slab is ground smooth and polished to remove saw marks. Usually grinding with abrasive (starting with grit size 220, then 400, then 600, and finally 5 micron) and water on a glass plate is sufficient. If the edges and top surface of a slab are coated with epoxy, the sample usually will not break apart during grinding and washing with water. Since water often causes the sample to disintegrate during this stage, the epoxy coating is essential. A lightweight household cooking or salad oil can be used in place of water as a lubricant. Dry grinding is possible by using abrasive papers of similar grit sizes. Finally, the rock is cleaned of all abrasive and polished with a 0.3 micron compound.

After polishing, the slab is cemented onto a standard petrographic slide using epoxy resin. Care should be taken to prevent bubbles from getting into the epoxy during mixing. The rock surface is coated and a slide placed over it. Bubbles are extruded by applying gentle pressure on the glass. This operation is best done by observing the sample under a light microscope. The epoxy resin should be cured according to manufacturers instructions. At this stage we usually coat the exposed top and sides of the rock slab with a thin epoxy layer to protect it from breaking during the next stages.

The sample is then cut down on a diamond slab saw or ground down by hand to about 1 mm. The final grinding procedure is similar to those previously described except that the hardened epoxy around the edge of the rock wafer on the slide should be removed (a fine diamond coated disc in a dental drill is helpful here) because it is removed at a slower rate than the rock itself during grinding, and slows down the grinding process. The sample should be checked periodically during grinding for proper thickness, recognized when quartz gives first order white interference color (at 30 µm thickness).

With patience and skill, a suitable thin-section for fabric analysis may be produced by this technique.

SCANNING ELECTRON MICROSCOPY

Microfabric is best observed using the scanning electron microscope (SEM). Observation of a macrofabric of a rock sample from an x-radiography photo and in thin-section allows one to determine which specific areas to study in detail with the SEM. Only a small sample (0.5mm^3) is needed for SEM viewing; however, this is usually prepared from a slightly bigger sample (e.g. 2-3 cm^2). First, a narrow groove is cut completely around the rock piece with a diamond coated thin grinding disc (see diagram). About 3 mm of sample is left uncut. Then, carefully place the pliers on both pieces and break at the cut to produce a freshly fractured surface. There will be a fine dust on the surface which should be blown free. This technique produces a fresh, unaltered surface. The sample is then shaped with the grinding disc and attached to an SEM stub, using a conducting (e.g. silver) cement or mounting tape. The sample is then either coated in a vacuum evaporator or sputter coated with a thin layer of conductive metal such as gold (used in our work). Typically, a thin line of silpaint is added by some investigators to provide an electrical ground from the sample to the stub. Applying the thin coat of conductive metal to the sample allows a clearer image to be obtained. Uncoated samples are very difficult to view or photograph. After coating the sample is ready for viewing. We suggest that before SEM viewing, the sample should be examined under a light microscope and a map made of various features on the small sample because it is easy to get lost when scanning at the higher magnifications of an SEM. SEM photomicrographs in this book were taken with an ISI-40 (International Scientific Instruments, Inc.) at 15 KV.

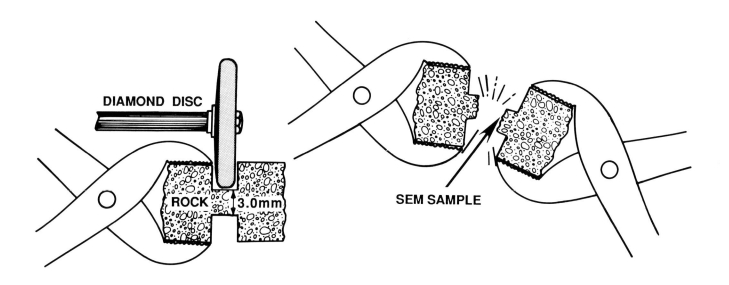

CHAPTER 3
X-RADIOGRAPHY, PETROGRAPHY
AND SCANNING ELECTRON MICROSCOPY DESCRIPTIONS

X-RADIOGRAPHY CLASSIFICATION OF ARGILLACEOUS ROCK MACROFABRIC

This section reveals three distinctive macrofabric types shown in x-radiographs of mudstones and shales. The classification is based upon the degree of development of lamination.

```
                    GROUP
DECREASING    ↓   A   WELL DEVELOPED LAMINATION
LAMINATION    ↓   B   INDISTINCTLY LAMINATED
                  C   BIOTURBATED
```

Significant work on categorizing lithic types of argillaceous rocks based upon macrofabric revealed by x-radiography has been published by Nuhfer et al. (1979), Cluff (1980), and Nuhfer (1981). Both Nuhfer and Cluff based their observations on fabrics of Devonian shale. We show here examples from various stratigraphic positions and sedimentary environments to illustrate (1) a macrofabric classification of argillaceous rocks, and (2) the usefulness of macrofabric in environmental interpretations. The scheme proposed here is intended to be simple and useful in making a quick and accurate macrofabric description and is based upon our observations and modification of the classifications used by Nuhfer and Cluff. All scale bars in photos equal 1 cm.

A. WELL DEVELOPED LAMINATION

Characteristics: Thin (<2mm; commonly <1mm) closely spaced laminae, commonly continuous across a sample, although they may be discontinuous in places. Alternating light and dark layers are due to an alternation of layers containing pyrite, silt, organics, and clay. Bioturbation is absent in samples with continuous layers and minor in sample containing discontinuous layering.

1. HURON SHALE MEMBER (Ohio Shale Formation, Upper Devonian), Lincoln County, West Virginia. Continuous laminae in the upper part; slightly discontinuous in the lower part.
2. MECCA QUARRY SHALE MEMBER (Linton Formation, Pennsylvanian), Park County, Indiana. Uninterrupted lamination.
3. BITUMINOUS SHALE FORMATION (U. Lias, Jurassic), Ravenscar, Yorkshire, England. Good example of continuous lamination uninterrupted by bioturbation or bottom currents.

GROUP A includes Cluff's "finely laminated" and "thickly laminated" shales and Nuhfer's examples of "thinly laminated," "lenticularly laminated" and "sharply banded."

B. INDISTINCT LAMINATION

Characteristics: Bedding or lamination is still visible, but disrupted. Some laminae continue across a sample and appear wispy. The macrofabric of this group results from slight disruption by bottom flowing currents and/or by bioturbation, but not as thorough as in the bioturbated macrofabric. Some burrows are visible.

1. CANTON SHALE MEMBER (Carbondale Formation, Middle Pennsylvanian) Gallatin County, Illinois. Although bioturbated in its central part, remnant lamination is still visible at the top and bottom. Sample is a gray shale.
2. CASHAQUA SHALE MEMBER (Sonyea Formation, Upper Devonian), Wyoming County, New York. This moderately bioturbated gray shale retains evidence of primary disrupted lamination. It exhibits poorly developed fissility in outcrop.
3. JAVA FORMATION (Upper Devonian) Lincoln County, West Virginia. Primary bedding is slightly disturbed by mild bioturbation. Thickness of laminae varies and laminae are discontinuous.

GROUP B includes Cluff's "indistinctly bedded" shale and Nuhfer's "non-banded" and "lenticularly laminated" shales.

C. BIOTURBATION

Characteristics: No bedding or lamination is apparent. A mottled pattern is visible. Burrows appear as dark, randomly-oriented lines on the x-radiograph. Bioturbation has destroyed all layered features.

1. GRAY SHALE IN FERRON SANDSTONE MEMBER (Mancos Shale Formation, Upper Cretaceous) Utah. The highly mottled appearance indicates extensive bioturbation, although there are no recognizable burrows.
2. CANTON SHALE MEMBER (Carbondale Formation, Middle Pennsylvanian) Gallatin County, Illinois. Dark areas are vertical and horizontal burrows. This sample is extensively bioturbated.
3. JAVA FORMATION (Upper Devonian) Jackson County, West Virginia. Extensive bioturbation is indicated by burrows (dark lines) and a mottled appearance.

GROUP C includes Cluff's "bioturbated mudstone" and Nuhfer's "non-banded" shale.

X-RADIOGRAPHY CLASSIFICATION OF ARGILLACEOUS ROCK MACROFABRIC

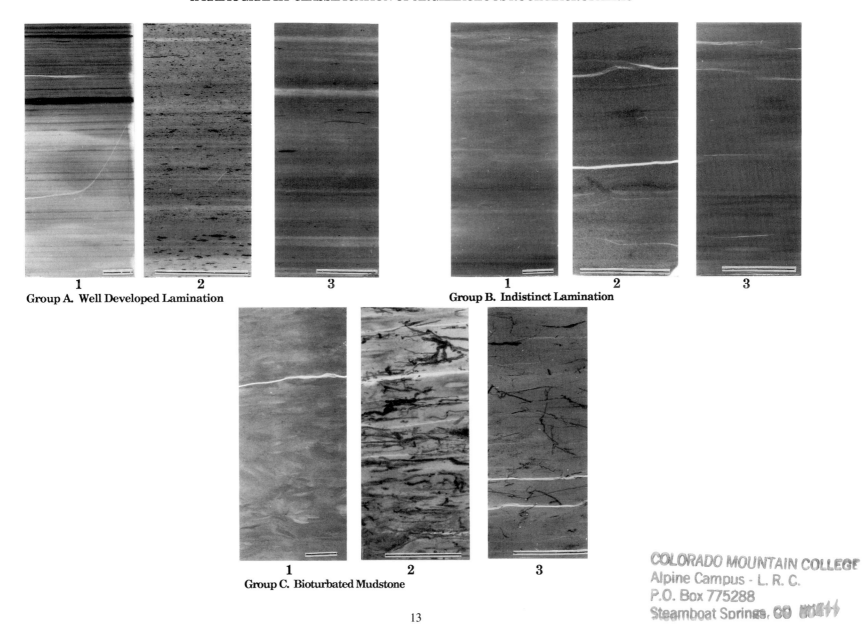

SUGGESTED ENVIRONMENTAL SIGNIFICANCE OF MACROFABRIC X-RADIOGRAPHY DATA

The figure on the next page shows the suggested environmental significance of macrofabric x-radiography data by relating macrofabric to sedimentary environments. The well developed lamination (Group A) is well preserved in the quiet deep density-stratified water (see pycnocline). Oxygen depletion in this zone restricts benthonic organisms thus minimizing sediment mixing by bioturbation. Dislocation of primary laminae is also prevented due to decreased velocity and/or complete absence of bottom flowing currents in the stagnant water. Indistinct lamination (Group B) represents deposition of sediment under slightly more oxygenated conditions which would allow some sediment mixing by those infaunal organisms that tolerate low oxygen conditions. In the diagram this zone is depicted as a transition from Anaerobic to Dysaerobic and is characterized in x-radiography by indistinct bedding. Alternately this macrofabric could result from slight reworking of sediment by bottom flowing currents in shallow water. The third fabric of Group C is definitely produced by biogenic activity. Bioturbation is recognized by the conspicuous burrows and greatly mottled texture. The original primary fabric is completely reworked to form bioturbated mudstone. Biogenic activity is promoted in moderately agitated to quiet shallow near shore water which contains a higher oxygen content from the other two offshore zones.

ANAEROBIC	ANAEROBIC-DYSAEROBIC	DYSAEROBIC-AEROBIC
(Quiet, deep water)	(Quiet water of moderate depth)	(Moderately agitated to Quiet, shallow water)
150 m - 200 m + (?)	± 150 m	< 150 ± m
WELL DEVELOPED LAMINATION	INDISTINCT LAMINATION	BIOTURBATED MUDSTONE

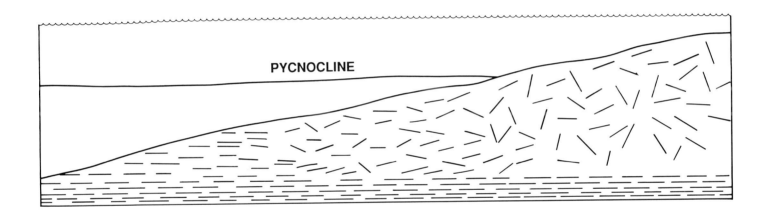

PETROGRAPHIC CLASSIFICATION OF BLACK SHALES

Organic-rich shales are commonly also called "black" or "bituminous" shales. Color ranges from dark gray to black depending upon the iron and/or carbon content. Hosterman and Whitlow (1983) compared color and organic carbon content of approximately 880 Devonian Shales from the Appalachian Basin. Carbon ranged from as low as 1% in medium gray to 2% or greater in dark gray to black shales. In addition to organic content, all black shales also share another common feature - lamination. Our petrographic analysis of over 100 dark gray to black Paleozoic and Mesozoic shales showed variations in the type of lamination present. This section illustrates the four main lamination types present in organic-rich shales: (1) finely laminated (2) thickly laminated (3) wavy laminated (4) lenticular laminated. Use of these macrofabric terms should facilitate a more accurate petrographic description of shales when combined with mineralogical, geochemical, and microfabric data. Photomicrographs here are representative of black shale macrofabrics observed in thin-section.

BLACK SHALE - LAMINATION TYPES

FINELY LAMINATED - alternating silt and organic-clay layers; layers are < 0.1 mm thick and exhibit parallel contacts. (A)

THICKLY LAMINATED - alternating silt and organic-clay layers; layers are > 0.1 mm thick and exhibit parallel contacts. (B)

WAVY LAMINATED - similar to Finely Laminated except the contacts are undulatory. (C)

LENTICULAR LAMINATED - possesses a "flaser bedding" appearance with lenses arranged in layers; lenses are composed of light colored (quartz, calcite) minerals encased by a dark (clay, organics) matrix. (D)

EXAMPLES

A. FINELY LAMINATED - GENESEO SHALE MEMBER
   (Genesee Formation, Upper Devonian)
   Ontario County, New York
   Scale = 0.1 mm
   Crossed nicols

B. THICKLY LAMINATED - HURON SHALE MEMBER
   (Ohio Shale Formation, Upper Devonian)
   Gallia County, Ohio
   Scale = 0.1 mm
   Plane light

C. WAVY LAMINATED - JET ROCK FORMATION
   (U. Lias, Jurassic)
   Port Mulgrave, Yorkshire, England
   Scale = 0.1 mm
   Plane light

D. LENTICULAR LAMINATED - ENERGY SHALE
   (Carbondale Formation, Pennsylvanian)
   Jefferson County, Illinois
   Scale = 0.1 mm
   Plane light

BLACK SHALES-LAMINATION TYPES

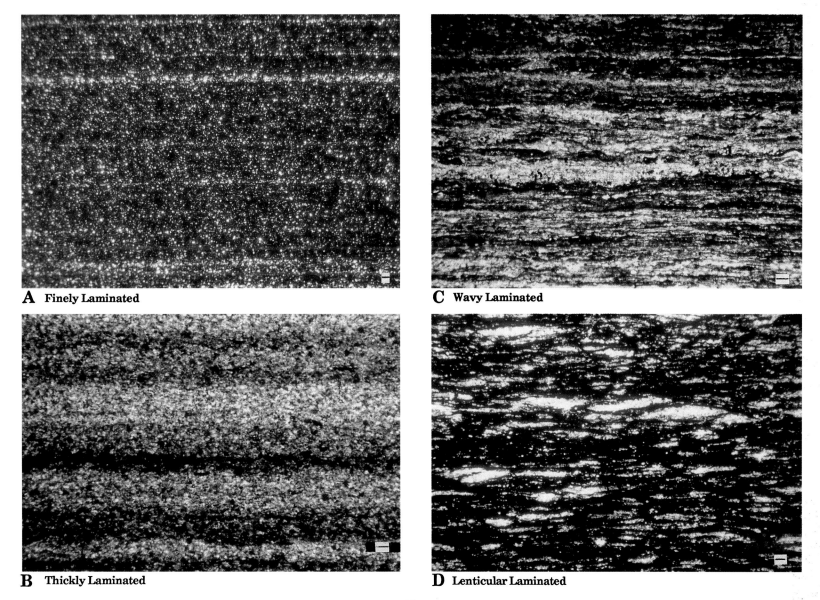

A Finely Laminated

B Thickly Laminated

C Wavy Laminated

D Lenticular Laminated

FINELY LAMINATED BLACK SHALES

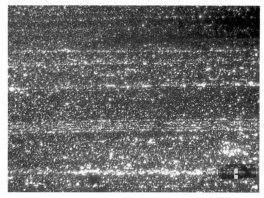 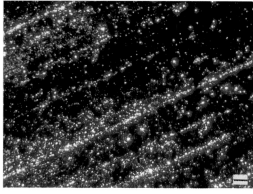 

GENESEO SHALE MEMBER (Genesee Formation, Devonian) Seneca County, New York. Fine layers of dark organic-rich material (organics plus clay) alternate with quartz silt layers (some only one to three grains thick). Scale = 0.1 mm. Crossed nicols.

PENN YAN SHALE MEMBER (Genesee Formation, Devonian) Livingston County, New York. Numerous silt layers, each one to two quartz grains thick. Scale = 0.1 mm. Crossed nicols.

GENESEO SHALE MEMBER (Genesee Formation Devonian) Ontario County, New York. Fine layers of dark organic-rich material (organics plus clay) alternate with quartz silt layers (some only one to three grains thick). Scale = 0.1mm. Crossed nicols.

 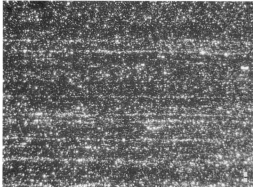 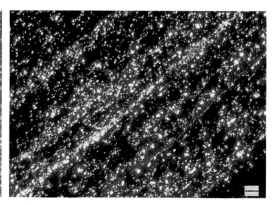

GENESEO SHALE MEMBER (Genesee Formation Devonian) Ontario County, New York. Scale = 0.1 mm. Plane light.

RHINESTREET SHALE MEMBER (West Falls Formation, Devonian) Livingston County, New York. Fine layers of dark organic rich materials (organics plus clay) alternate with quartz silt layers (some only one to three grains thick) (similar to A and B). Scale = 0.1 mm. Crossed nicols.

UTICA SHALE (Ordovician) Jefferson County, New York. Silt layers < 0.1 mm thick alternate with clay and organic layers. Scale = 0.1 mm. Crossed nicols.

THICKLY LAMINATED BLACK SHALES

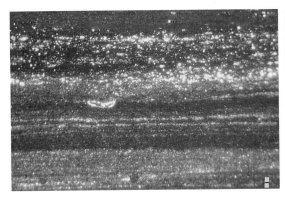

CASHAQUA SHALE MEMBER (Sonyea Formation, Devonian) Livingston County, New York. The lower part of this shale is mildly bioturbated, however, the upper part exhibits thick laminations undisturbed by mixing. Scale = 0.1 mm. Crossed nicols.

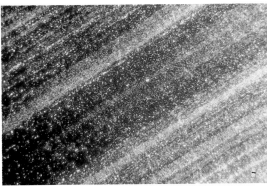

MIDDLESEX SHALE MEMBER (Sonyea Formation, Devonian) Ontario County, New York. Alternation of dark organic-rich layers with lighter colored silty layers. Scale = 0.1 mm. Crossed nicols.

HURON SHALE MEMBER (Ohio Shale Formation, Devonian) Gallia County, Ohio. Sample shows banding due to variation in organic concentration. Abundant flattened spore cases are aligned parallel to laminae. Scale = 0.1mm. Plane Light.

CHAGRIN SHALE MEMBER (Ohio Shale Formation, Devonian) Gallia County, Ohio. Alternation of light colored silt-rich layers with darker colored organic rich layers. Scale = 0.1 mm. Crossed nicols.

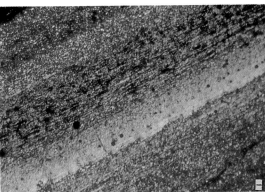

JET ROCK FORMATION (U. Lias, Jurassic) Port Mulgrave, Yorkshire, England. Organic-rich layers (dark) alternating with silt- and clay-rich layers (lighter). Scale = 0.1 mm. Crossed nicols.

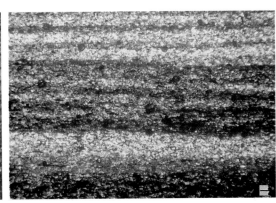

HURON SHALE MEMBER (Ohio Shale Formation, Devonian) Gallia County, Ohio. Alternation of light colored silt-rich layers with darker colored organic-rich layers. Scale = 0.1 mm. Plane light.

WAVY LAMINATED BLACK SHALES

JET ROCK FORMATION (U. Lias, Jurassic) Port Mulgrave, Yorkshire, England. Samples of Jet Rock exhibit a "stromatolitic-type" of lamination of alternating wavy dark organic-rich layers with lighter colored silt-clay laminae. Scale = 0.1 mm. Plane light.

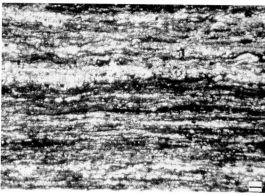

JET ROCK FORMATION (U. Lias, Jurassic) Port Mulgrave, Yorkshire, England. Abundant light colored silt (quartz?) layers and dark colored organic-clay rich layers. Scale = 0.1 mm. Plane light.

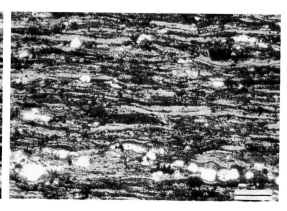

WHITE BAND - "PECTINATUS ZONE" (Kimmeridge Clay Formation, Jurassic) Kimmeridge Bay, Dorset, England. Compressed marine algae spore cases (yellow) are apparent. Quartz grains appear as clear areas. Scale = 0.1mm. Plane Light.

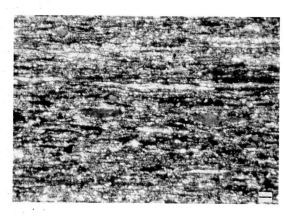

BITUMINOUS SHALE (U. Lias, Jurassic) Ravenscar, Yorkshire, England. Silt, organic, and clay layers. Scale = 0.1 mm. Plane light.

BLACKSTONE LAYER (Kimmeridge Clay Formation, Jurassic) Kimmeridge Bay, Dorset, England. Compressed marine algae spore cases (yellow) are apparent. Scale = 0.1 mm. Plane light.

CLEVELAND SHALE MEMBER (Ohio Shale Formation, Devonian) Gallia County, Ohio. Compressed marine algae spore cases (yellow) are apparent. Layers of coarse silt size grains alternate with darker colored organic and clay layers. Scale = 0.1 mm. Plane light.

LENTICULAR LAMINATED BLACK SHALES

ENERGY SHALE MEMBER (Carbondale Formation Pennsylvanian) Jefferson County, Illinois. Lenticular lamination characterizes the organic-rich Pennsylvanian age black shales of the United States Midcontinent region. Scale = 0.1 mm. Plane light.

MECCA QUARRY MEMBER (Linton Formation, Pennsylvanian) Parke County, Indiana. Scale = 0.1 mm. Plane light.

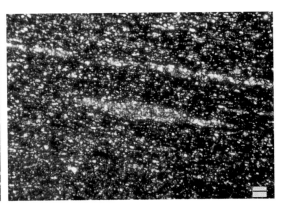

UPPER CLEGG CREEK MEMBER (New Albany Formation, Devonian) Clark County Indiana. Elongate lens of quartz grains (center of photo). Scale = 0.1mm. Crossed nicols.

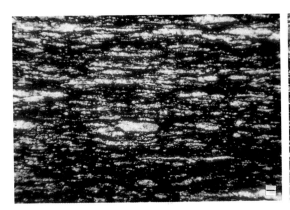

HUSHPUCKNEY SHALE MEMBER (Swope Formation, Pennsylvanian) Adair County, Iowa. This shale is not associated with a coal but with gray shale and limestone; however, it still possesses characteristic lenticular bedding. Scale = 0.1 mm. Plane light.

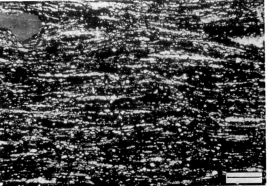

ROOF SHALE OVER SPRINGFIELD V COAL (Petersburg Formation Pennsylvanian) Vanderburg County, Indiana. Scale = 0.1 mm. Plane light.

ROOF SHALE OVER SPRINGFIELD V COAL (Petersburg Formation Pennsylvanian) Knox County, Indiana. Scale = 0.1 mm. Plane light.

VARIOUS PETROGRAPHIC FEATURES OF SHALE - BIOTURBATION

WINDOM SHALE MEMBER (Moscow Formation, Devonian) Livingston County, New York. This gray shale has poorly developed fissility in outcrop and is very fossiliferous. Notice the lack of lamination in the thin-section. Scale = 0.1 mm. Plane light.

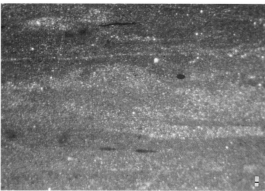

CASHAQUA SHALE MEMBER (Sonyea Formation, Devonian) Wyoming County, New York. This green shale has moderate to good fissility in outcrop. The thin section exhibits disruption of laminae due to moderate bioturbation. Scale = 0.1 mm. Crossed nicols.

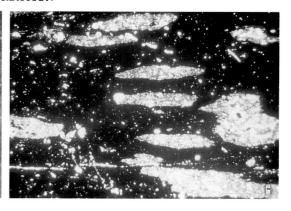

TURNER MINE SHALE MEMBER (Carbondale Formation, Pennsylvanian) Gallatin County, Illinois. Normally most black shales are well laminated. The Chondrites trace fossil burrows (light areas) in this black shale indicate post-depositional bioturbation. Extensive bioturbation destroyed lamination. Scale = 0.1 mm. Plane light.

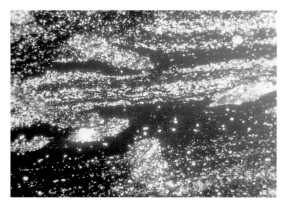

GRAY SHALE FORMATION (U. Lias, Jurassic) Port Mulgrave, Yorkshire, England. This thin-section shows a coarser material filling burrows and the swirled nature of lamination. Burrows are the lighter areas. Obviously this rock was originally laminated and laminae were disrupted during bioturbation. Scale = 0.1 mm. Plane light.

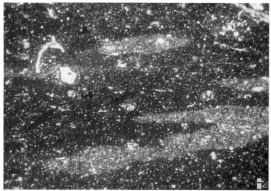

GRAY SHALE FORMATION (Lias, Jurassic) Port Mulgrave, Yorkshire, England. This sample shows coarser material filling burrows and the swirled nature of lamination. The burrows are the lighter colored areas. Scale = 0.1 mm. Plane light.

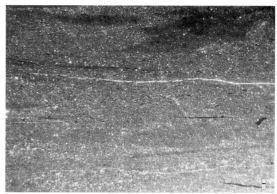

WEST RIVER SHALE MEMBER (Genesee Formation, Devonian) Ontario County, New York. This dark gray shale is fissile in outcrop, but still has zones which exhibit bioturbation. The thin section is from a moderately mixed zone. Notice disruption of lamination in the upper part. Scale = 0.1mm. Plane light.

PETROGRAPHIC FEATURES OF SHALE - MISCELLANEOUS

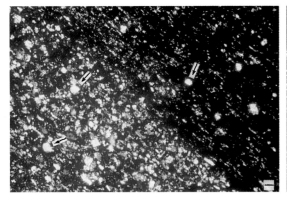

JET ROCK FORMATION (U. Lias, Jurassic) Ravenscar, Yorkshire, England. Arrows point to silt-size quartz grains. Notice the sharp contact between the silt-abundant layer (lower left) and the overlying darker organic and clay-rich layer. Scale = 0.1 mm. Crossed nicols.

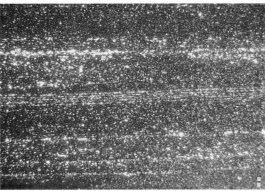

GENESEO SHALE MEMBER (Genesee Formation, Devonian) Seneca County, New York. Five separate silt laminae occur in the center of the thin-section, each composed of a single layer of quartz grains. Single-grain laminae are common in Devonian black shales of New York. Scale = 0.1 mm. Crossed nicols.

PENN YAN SHALE MEMBER (Genesee Formation Devonian) Ontario County, New York. Organic-rich, dark gray shale displaying micro "graded bedding" in the silt-clay size range. The arrow indicates the top of the bed. Scale = 0.1mm. Plane light.

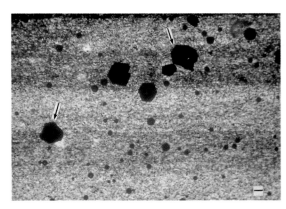

CHAGRIN SHALE MEMBER (Ohio Shale Formation, Devonian) Gallia County, Ohio. Pyrite appears opaque in this thin-section of dark gray shale (arrows). Notice the extreme variation in sizes of pyrite crystals. Scale = 0.1 mm. Crossed nicols.

PENN YAN SHALE MEMBER (Genesee Formation, Devonian) Ontario County, New York. Micro-crossbedding in a black shale. Notice the variation in the thickness of light colored silt laminae. Scale = 0.1 mm. Crossed nicols.

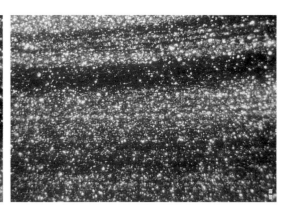

PENN YAN SHALE MEMBER (Genesee Formation, Devonian) Ontario County, New York. Fine scale current bedding occurs at the top of this thin-section. Scale = 0.1 mm. Crossed nicols.

SCANNING ELECTRON MICROSCOPY DESCRIPTIONS
PREFERRED PARTICLE ORIENTATION IN SHALE

Preferred particle orientation (parallelism of platy minerals, mainly clay minerals) is a characteristic of most non-bioturbated shales, especially dark gray to black shales. The latter are usually not bioturbated and hence the original primary rock fabric is well preserved. When did this primary fabric form? Did the parallelism result from deposition of individual dispersed clay flakes? Or were the randomly oriented platelets in an original flocculated clay sediment reoriented into a parallel position soon after deposition and during burial? Either mechanism could produce the same final fabric as the adjacent figures show.

This section illustrates the fabrics of laboratory controlled clay deposition and three typical shales to show well-developed preferred particle orientation. Silt-sized grains are absent in these samples.

Two possible mechanisms for producing primary shale fabric from flocculated or dispersed clay are shown below. Models are based upon a sediment with high clay content. Observations indicate that the silt content also has an influence on argillaceous rock fabric. The greater amount of silt, the more random is the fabric.

A. DISPERSED ILLITE - Sample is grundite illite mixed with a dispersing agent in distilled water, settled in a glass tube, and air dried. This example is offered to illustrate the final orientation produced in a sediment by dispersed clay. Parallelism of particles is extremely well developed and compares favorably with the fabric of some shales. Scale = 10 μm.

B. BITUMINOUS SHALE FORMATION (Upper Lias, Jurassic) Ravenscar, Yorkshire, England. This organic rich, fissile marine shale is described elsewhere in the Atlas. The remarkable development of preferred orientation of clay is similar to that of dispersed illite. Can clay settle in the dispersed state in electrolyte-rich marine water or does it always flocculate? On the other hand, does a high organic material content promote dispersion? Two models are proposed in the diagram for the origin of preferred orientation. Model 1 indicates that clay flakes arranged in a cardhouse fashion in flocculated sediment reorient upon early burial and compaction. Model 2 shows parallelism resulting from deposition of dispersed clay. Scale = 10 μm.

C. BEDFORD FORMATION (Mississippian) Gallia County, Ohio. Preferred orientation in this shale is apparent. Notice how clay flakes wrap around the silt (quartz?) grain in the left center of the photo. Scale = 10 μm.

D. SUNBURY FORMATION (Mississippian) Gallia County, Ohio. This shale exhibits well developed parallelism of platy material. Scale = 10 μm.

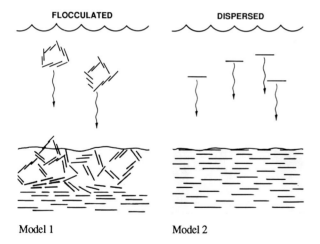

Model 1   Model 2

SCANNING ELECTRON MICROSCOPY DESCRIPTIONS-PREFERRED PARTICLE ORIENTATION IN SHALE

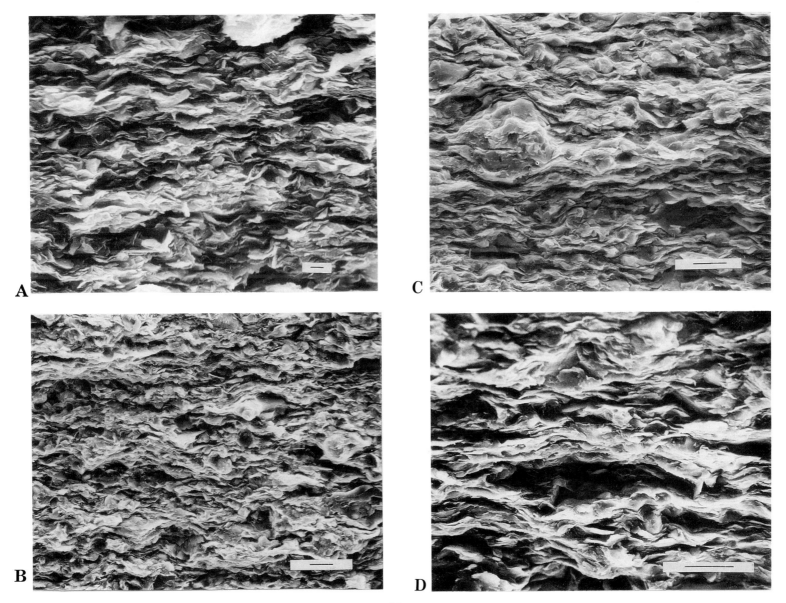

RANDOM PARTICLE ORIENTATION IN MUDSTONE

Random particle orientation is characteristic of mudstones or non-fissile argillaceous rocks. This particular arrangement may be produced by preservation of the primary fabric of flocculated clay or secondarily produced by bioturbation. The diagram below shows several commonly accepted models of flocculated clay (after Van Olphen, 1963 and Bennett et al., 1979). Observations by us and by Bennett et al. (1979) of flocculated clays reveal a fabric dominated by randomly oriented domains of stepped face-to-face oriented flakes in each domain. Domains, in turn, are randomly oriented in a typical flocculated cardhouse structure. Also found in argillaceous rocks and sediment may be a random orientation of individual flakes arranged edge-face and face-face. Thus, it seems that various orientations exist in a flocculated clay mass, however, domains do tend to dominate. These fabrics are shown in panels A, B and C. The bioturbated fabric in D also is random, but it is characterized by the dominance of randomly oriented individual flakes rather than domains or clusters. This observation is significant in differentiating the origin of the fabric.

MODELS OF FLOCCULATED CLAY

DOMAINS   EDGE-FACE

FACE-FACE   EDGE-EDGE

A. FLOCCULATED ILLITE - This stereo micrograph shows the fabric of grundite illite (1 gm/l) flocculated in salt water (27 NaCl gm/l), sedimented in a glass tube, and air-dried. Randomness of domains plus some individual flakes prevails in this sample. Compare this fabric to the models of flocculated clay shown above and to natural sediment and rock fabrics in SEM micrographs B and C. Scale = 1 μm.

B. GREAT SALT LAKE SEDIMENT (Pleistocene) Great Salt Lake, Utah. Fabric in a freeze-dried core from Great Salt Lake. This sample shows no evidence of bioturbation in hand specimen. Randomness of flakes is dominant. A few larger silt grains are present. Notice the porous fabric similar to the flocculated clay in A. The high salinity of the depositional environment of this clay is a major factor in producing the random fabric of this flocculated clay. Scale = 10 μm.

C. CLEAR FORK FORMATION (Permian) Swisher County, Texas. This mudstone was deposited in a saline mudflat associated with other Permian evaporite deposits. The sample is not bioturbated and it is sandwiched between thin halite layers, thus indicating the clay was deposited in water high in electrolytes and hence flocculated. Domains are abundant. Scale = 1 μm. Used with permission SEPM (Society for Sedimentary Geology).

D. CASHAQUA MEMBER (Sonyea Formation, Devonian) Wyoming County, New York. X-radiography and outcrop viewing provide abundant evidence of activity by organisms which modified the original fabric of the Cashaqua sediment. This SEM micrograph shows a random mixture of silt and individual clay platelets which is typical of bioturbated fabric. Scale = 10 μm.

References:
Bennett et al. (1979), Van Olphen (1963), O'Brien (1987).

SCANNING ELECTRON MICROSCOPY DESCRIPTIONS-RANDOM PARTICLE ORIENTATION IN MUDSTONE

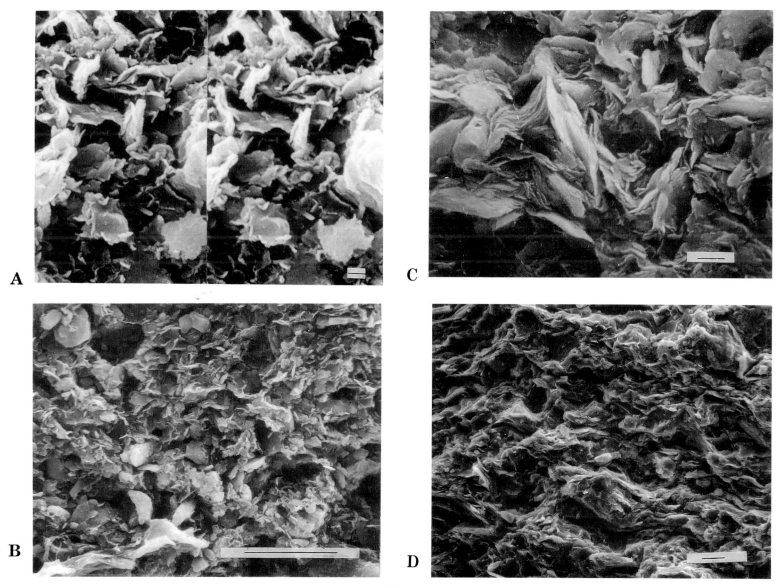

27

FABRIC VARIATIONS IN ORGANIC RICH SHALES

Even though preferred orientation dominates, organic rich shale microfabrics revealed by the SEM do vary. This section illustrates four main types of microfabric found in an SEM analysis of over 100 organic rich shales from various stratigraphic positions and sedimentary environments. The fabric nomenclature is based upon a visual estimate of fabric variation which, in turn, is related to abundance of clay, particulate organic matter, and silt grains. The latter two materials greatly influence microfabric of these shales. Fabric types are: (A) ORGANIC HASH, (B) ORGANIC-CLAYEY, (C) CLAYEY, (D) SILTY. For example, in describing an organic rich shale under the SEM, one might refer to it as composed of "organic hash" or as an "organic-clayey black shale." There is no attempt here to propose an organic shale classification, but only to clarify analysis of organic rich shales by supplying suitable descriptive terminology. This analysis may be of value to those undertaking studies of hydrocarbon production potential of organic rich shales.

MICROFABRIC TYPES OF ORGANIC RICH SHALES

ORGANIC HASH - composed dominantly of fine (<5 μm) discrete organic particles preferentially oriented with platy minerals. Organics dominate; their edges are diffuse. Example A.

ORGANIC-CLAYEY - resembles ORGANIC HASH except platy minerals are recognized by their well-defined shape and are more visible. Example B.

CLAYEY - composed dominantly of discrete well-defined platy clay minerals oriented parallel to bedding. Silt (quartz) grains are a minor component. Example C.

SILTY - silt grains (c. 10-20 μm) are very visible in a clay matrix. Particle orientation becomes less preferred with increasing silt content. Example D.

References:
    Bitterli (1963), O'Brien (1968), Spears (1976), Curtis (1980), Ece (1987).

A. ORGANIC HASH - MECCA QUARRY SHALE MEMBER (Linton Formation, Pennsylvanian) Park County, Indiana. TOC = 49.3%. Scale = 10 μm.

B. ORGANIC-CLAYEY - ROOF SHALE SPRINGFIELD COAL MEMBER (Petersburg Formation, Pennsylvanian) Vanderburg County, Indiana. TOC = 26.2%. Scale = 10 μm.

C. CLAYEY - GENESEO MEMBER (Genesee Formation, Upper Devonian) Seneca County, New York. TOC = 1.7%. Scale = 10 μm.

D. SILTY-UTICA SHALE (Ordovician) Jefferson County, New York. TOC = 3.8%. Scale = 10 μm.

SCANNING ELECTRON MICROSCOPY DESCRIPTIONS OF BLACK SHALE FABRIC VARIATIONS

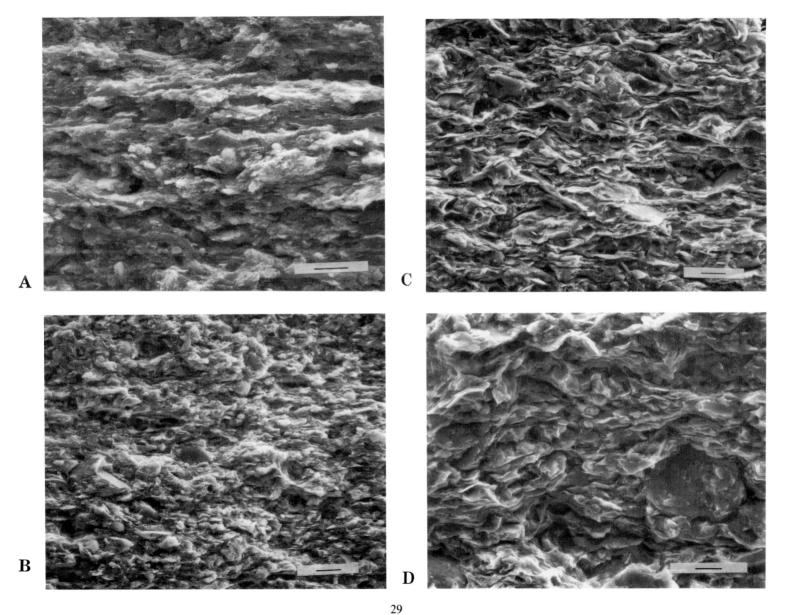

CHAPTER 4
MISCELLANEOUS FEATURES IN ARGILLACEOUS ROCKS

PYRITE FRAMBOIDS

"Framboid" describes spheroidal aggregates of pyrite microcrystallites (Rickard, 1970). They are common in organic rich shales and give clues to geochemical and microfabric conditions. Rickard indicates the spheroidal shape may be due to pseudomorphism of a pre-existing spherical body. It may represent replacement of an immiscible organic liquid globule by iron sulfide or infilling by microcrystalline pyrite of spherical gaseous vacuoles or even by pseudomorphism of single-celled microorganisms. Framboid morphology is also attributed to pyritization of microflora (Javor and Mountjoy, 1976). Amstutz et al. (1967) indicate the framboid is inherited from the colloidal glob of iron monosulfide formed authigenically in sediment. Authigenic pyrite is found with organic matter and is an indicator of anaerobic sulfide diagenesis (Berner, 1970). Field and laboratory observations (Berner, 1969; Love, 1967) show pyrite forming at shallow sediment burial depth (10 cm to 3 m) under low temperature (<80° C). A good brief summary of their origin is given by Kalliokoski (1974). Observations suggest framboid formation in wet, unconsolidated sediment, probably associated with a flocculated clay of soupy consistency. Thus, their presence offers clues to the early burial diagenesis history of clayey sediment.

FRAMBOID FORMATION

1. GAS BUBBLE (H2S?, CO2?) FORMS IN SOFT FLOCCULATED CLAY SEDIMENT

2. ADJACENT CLAY ORIENTS AROUND GAS BUBBLE AS IT GROWS IN SOUPY, EASILY DEFORMED, FLOCCULATED SEDIMENT

3. AUTHIGENIC PYRITE FILLS GAS BUBBLE CAVITY, PARTICLE ORIENTATION IS STILL RANDOM

4. OVERBURDEN PRESSURE REORIENTS REMAINING CLAY INTO PARALLEL POSITION COMMON IN SHALE

* models modified after Rickard (1970, p. 289)

AUTHIGENIC PYRITE MICROCRYSTALLITES

A$_1$. HUSHPUCKNEY MEMBER (Swope Formation, Pennsylvanian) Adair County, Iowa. The pyrite octahedra are found in small depressions in this organic rich marine shale. TOC = 19.3%. Scale = 1 μm.

A$_2$. TYPICAL PYRITE FRAMBOID SPHERICAL SHAPE-ENERGY SHALE MEMBER (Carbondale Formation, Pennsylvanian) Jefferson County, Ilinois. The micrograph shows a mosaic of pyrite microcrystallites in a typical spherical framboid. Scale = 1 μm.

B. BITUMINOUS SHALE FORMATION (Upper Lias, Jurassic) Ravenscar, Yorkshire, England. Two framboid clusters have crystallized within a cavity. Note how the clay flakes closest to the framboid are oriented tangential to the surface. This orientation is probably produced in the manner shown in the adjacent diagrams. Scale = 10 μm.

C. RHINESTREET SHALE MEMBER (West Falls Formation, Devonian) Astabula County, Ohio. A typical example showing the well developed spherical nature of a framboid containing numerous pyrite crystallites. Notice platy particle orientation around the framboid. Compare to figure 4 in adjacent panel. Scale = 10 μm.

D. ENERGY SHALE MEMBER (Carbondale Formation, Pennsylvanian) Jefferson County, Illinois. This polyframboid has an ellipsoidal shape probably due to deformation during compaction. Preferred clay orientation in this shale is a result (see figure 4) of clay reorientation. The large polyframboid (diameter = 150 μm) probably developed during early stages of diagenesis when moist sediment was being compacted to form shale fabric. Deformation and growth of the large polyframboid and clay particle reorientation were probably occurring simultaneously. Because of their size (e.g. diameter = c. 10 μm), smaller individual spherical framboids resisted deformation during the formation of shale fabric. Scale = 10 μm.

PYRITE FRAMBOIDS

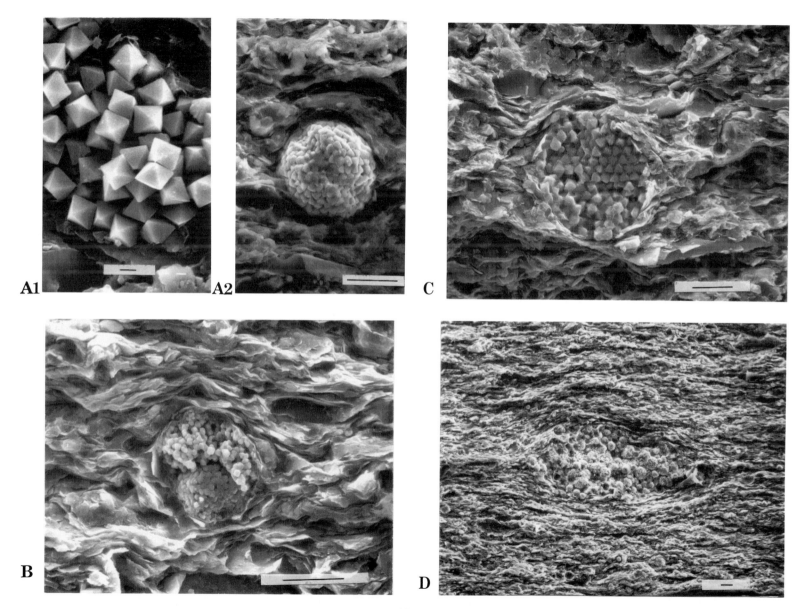

FECAL PELLETS

This section illustrates another microfabric feature of argillaceous rocks - fecal pellets. Although not common, when present in shale they may be recognized because their more random fabric contrasts to that of the enclosing shale. Porter (1984) states that few reports of microscopic (<0.5 μm) fossil fecal pellets appear in the literature. However, biogenic pelletization may be the most important process in depositing argillaceous sediments in shallow marine interdeltaic environments (Pryor, 1975). Preservation of fecal pellets in the rock record is favored by anoxic conditions which eliminate or minimize both biological and chemical degradation. Shown here are typical examples seen in SEM of fossil fecal pellets found in argillaceous rocks which are compared to the morphology of recent pellets. Note the conspicuous presence of phytoplanktonic matter (diatoms and coccoliths) mixed with lithogenic material in both recent and ancient examples. Note also the similarity of recent and ancient pellets in size and shape in microfabric. The high concentration of easily identified phytoplanktonic matter in a random mixture of clay and silt grains is a distinguishing feature of pellets. Hattin (1975) found coccolith-rich fecal pellets in Cretaceous rocks which were produced by copepods and tunicates. Copepods were also found to pelletize inorganic sediments as they fed on suspended sediment (Smith and Syvitski, 1982). Thus, investigators should be aware that the fabric analysis of argillaceous rocks (probably the more organic rich samples) may reveal fossil fecal pellets which may be useful in interpreting sedimentary conditions.

A. FECAL PELLET (Recent) Newfoundland Slope, Atlantic Ocean, 49°45.1'N, 49°26.2'W, depth 1400 m. This fecal pellet of unknown organism is composed of a random mix of lithogenic (quartz?) and biogenic (coccoliths, diatoms-small arrow; see small area, inset) material. Scale = 10 μm.
The inset micrograph shows a magnified view of the pellet (see large arrow for location of inset). Scale = 10 μm.

B. FECAL PELLET (Recent) Emerald Basin, Atlantic Ocean, 44°N, 63°W, off Nova Scotia, depth 250 m. Left SEM micrograph shows two halves of a pellet (350 μm x 150 μm) broken open in the lab (Scale = 100 μm).
A close up view on the right reveals a random orientation of platy material (clay), lithogenic material, and diatom fragments. Scale = 10 μm.

C. BLACK SHALE ROOF ROCK OF HOUCHIN CREEK COAL (Pennsylvanian) Gibson County, Indiana. Random orientation of fine material in a pellet located in center of photo. The morphology and size (50 μm x 25 μm) is similar to that of recent pellets. Notice the preferred particle orientation in the enclosing black shale which contrasts to the fabric of the pellet. Scale = 10 μm.

D. BITUMINOUS SHALE FORMATION (U. Lias, Jurassic) Ravenscar, Yorkshire, England. The size and oblong shape of this pellet are comparable to the Recent pellets. The oblong shape of this pellet is in part due to post-depositional compaction of wet flocculated sediment which reoriented to produce the preferred orientation of the enclosing shale. Scale = 10 μm.
Coccoliths (inset) are the dominant components of the pellet. Scale = 1 μm.

FECAL PELLETS

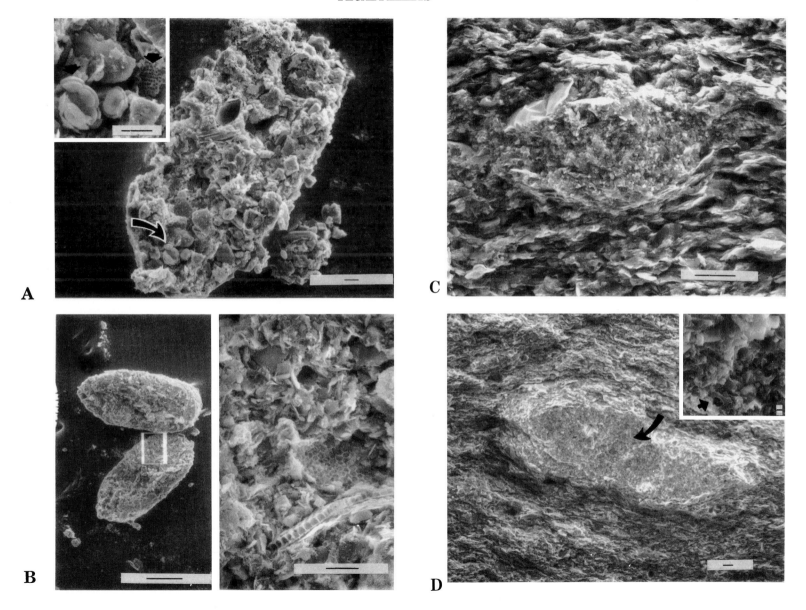

PALYNOMORPHS IN SHALES

Microscopic resistant-walled organic bodies which include pollen and spores are called palynomorphs (Traverse, 1988) and may occur in shales. Other palynomorph bodies include acritarchs, dinoflagellate thecae and cysts, certain colonial algae, scolecodonts, chitinozoan, and other acid-insoluble microfossils (Traverse, 1988). Yellowish marine algal bodies are visible in examples shown here of some dark gray to black shales. One spore-like body, *Tasmanites*, is very commonly found in Devonian - Mississippian dark shales. *Tasmanites* or *Tasmanites*-like bodies are planktonic algae and their presence suggests that pure tasmanite deposits accumulated from algae blooms (Tschudy and Scott, 1969). Other spores are from land plants and provide evidence of vegetation type adjacent to the sedimentary basin. Only by using techniques to extract them from the rock is it possible to study them in sufficient detail to make a positive genus identification. In thin-section, we have found the spore-like bodies are compressed. Their original shape is commonly spherical. A thin-section cut parallel to bedding reveals a circular shape, however, the appearance in a section perpendicular to bedding is that of a "deflated basketball." A compression ratio of 5 to 1 is calculated for these bodies This ratio is conservatively used to indicate the minimum amount of compaction that took place during burial and adds further evidence supporting the conclusion that the shale-forming sediment was initially flocculated. This unit illustrates palynomorph morphology in thin-sections and SEM of shales.

All samples shown here are from black shale.
In SEM photographs of surfaces perpendicular to bedding, spores have a distinctive "worm-like" morphology, distinct from surrounding organic matter.

A. CLEVELAND SHALE MEMBER (Ohio Shale Formation, Devonian) Gallia County, Ohio. Cross sections of deformed yellowish algal bodies are in the center of the photo. The thickness of the spore wall of the largest body is approximately 0.01 mm. Silt size quartz grains are light gray. Dark kerogen and clay is visible throughout. Scale = 0.1 mm. Plane light.

B. MARCELLUS SHALE FORMATION (Devonian) Mason County, West Virginia. The left photo is a view looking down onto a bedding plane surface and reveals the circular (originally spherical shape) of *Tasmanites*(?) spore-like bodies. A cross-section view in the right photo reveals spores. Notice that most do not show both spore walls, as seen in the Cleveland shale example. Scale = 0.1 mm. Plane light.

C. HURON SHALE MEMBER (Ohio Shale Formation, Devonian) Jackson County, West Virginia. The greatly flattened spore in the lower half of the SEM photo originally had a diameter of approximately 200 µm and offers evidence of compaction of the sediment enclosing it. This sediment was originally flocculated and reoriented upon burial during which the spore was compressed. Scale = 10 µm. Plane light.

D. CLEVELAND SHALE MEMBER (Ohio Shale Formation, Devonian) Gallia County, Ohio. This is an SEM photo of a *Tasmanites-like* (?) body. No cell wall is visible. Note how compaction has distorted the shape (compare to photo C). Scale = 10 µm. Plane light.

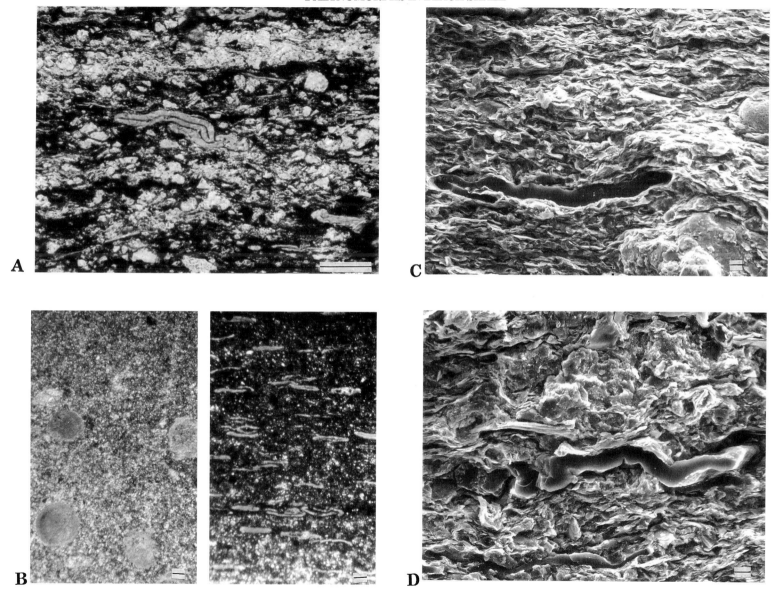

CHAPTER 5
CASE STUDIES OF SPECIFIC DISTINCTIVE FEATURES

WELL DEVELOPED LAMINATION IN A BLACK SHALE (EXAMPLE I)

BITUMINOUS SHALE FORMATION (Upper Lias, Jurassic) Ravenscar, Yorkshire, England.

Organic-rich shales are commonly well laminated and exhibit a microfabric of parallel clay flakes. This unit illustrates a sedimentological interpretation of a black shale possessing well developed lamination and preferred particle orientation. The Bituminous Shale outcrops along the northeast Yorkshire coast of England. This fissile, organic shale is in a classic area for the study of Jurassic ammonites.

Sedimentary Environment: marine, anaerobic
Significant Features: lamination, preferred orientation, organic content
Geology: This shale is characterized by a sparce, low diversity epifauna (mainly bivalves) (Morris, 1979, 1980). The macro- and microfabrics are characterized by well preserved laminae and parallelism of particles indicating an absence of bioturbation. The organic content indicates preservation of organics under anoxic conditions. Hallam (1967, 1975) has suggested that Jurassic shales were deposited slowly (approximately 0.3 mm/yr.) in quite shallow water ("no more than a few tens of metres deep"). Preservation of the thin laminae indicates lack of agitation by strong bottom currents after deposition.

Whole Rock Composition	Wt.%
Quartz	20
Plagioclase Feldspar	2
Calcite	14
Dolomite	1
Pyrite	11
Layer Silicates	49*
Illite-Smectite	31
Illite-Mica	37
Kaolinite	25
Chlorite	7
TOC (Total Organic Carbon)	3.1

Note: a detailed geochemical analysis of Jet Rock Series (including Bituminous Shale) is also found in Gad et al. (1969).

*49% of the whole rock is comprised of layer silicates, when considered as 100% of the clay fraction, the layer silicates are comprised of 31% illite-smectite, 37% illite-mica (discrete illite), 25% kaolinite, and 7% chlorite.

A. The x-radiograph shows well developed, alternating silt and organic - clay rich layers. Thickness of the laminae varies from 2.0 mm to 0.4 mm. Laminae are continuous across the sample. A detailed analysis of a lamina couplet is presented in a later section. Scale = 1 cm. Reprinted with permission from O'Brien, 1990.

B. Particulate organic matter appears black in a thin-section of this thickly laminated shale. Platy organic fragments lie parallel to bedding and obviously contribute to the fissile character of the rock. Vertical variation in color in thin-section is an indication of variation in organic concentration. Clear areas are quartz grains, darker areas are clay or organics. Scale = 0.1 mm. Plane light.

C. An SEM of this "clayey" black shale emphasizes the dominant preferred orientation of platy material (it is difficult to differentiate weathered clay from platy organic fragments in SEM). Scattered throughout are fine silt-size quartz grains. Cavities are zones where silt grains were dislodged during sample preparation. Scale = 10 μm.

WELL DEVELOPED LAMINATION IN A BLACK SHALE (EXAMPLE I)

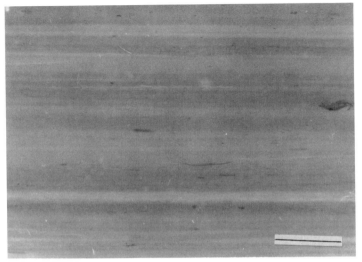

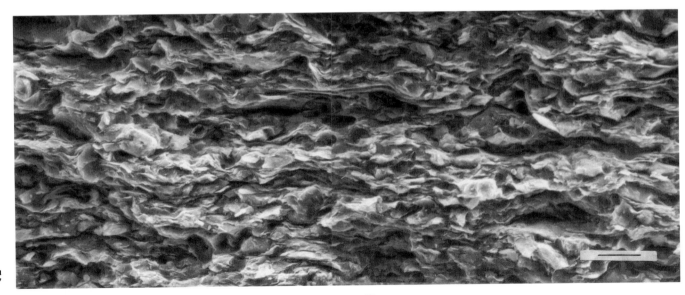

WELL DEVELOPED LAMINATION IN A BLACK SHALE (EXAMPLE II)

JET ROCK SHALE FORMATION (Upper Lias, Jurassic) Port Mulgrave, Yorkshire, England.

This example also illustrates well developed lamination in organic-rich black shale.

Sedimentary Environment: anaerobic, marine

Significant Features: thin, well developed, and wavy lamination.

Geology: The Jet Rock Shale is a well known shale underlying the Bituminous Shale. The original lamination is preserved because of the lack of bioturbation due to an entirely epifaunal group of organisms of very low diversity (Morris 1980). SEM micrographs show a repetition of fine organic and clay-rich laminae alternating with coarser silty layers suggestive of cyclic sedimentation. The high organic content (TOC=7.8%) indicates anoxic conditions at or near the bottom. A quiet water stagnant condition seems to have prevailed, however, it was interrupted periodically by deposition of silt. The fabric shown here is also similar to that of the Rundle Oil shale discussed in the section on Hydrocarbon Source Rock. A microbial mat origin is offered for the development of the lamination characteristic of the Rundle shale. Fabric similarity suggests this is a reasonable explanation for the origin of the wavy lamination of the Jet Rock shale.

A. The x-radiograph shows alternating organic and silt/clay rich layers. This would be classified as well developed lamination. Average lamina thickness varies from 0.1 - 0.4 mm. Scale = 1 cm. Reprinted with permission from O'Brien, 1990.

B. Thin-section (plane light) shows dark colored organic and clay rich layers alternating with lighter colored quartz silt layers. Wavy lamination is well displayed. Scale = 0.1 mm.

C. An SEM of the microfabric of a lighter colored silt layer. In the right-center (arrow) is a clay-encrusted silt grain (quartz). Notice parallelism of platy material (clay flakes). This orientation probably was produced upon sediment compaction by the reorientation of flocculated clay. Scale = 10 μm.

D. SEM of the microfabric in an organic/clay-rich lamina. Notice the high degree of parallelism of material due to compaction. The SEM description of this black shale would be "silty black shale." Scale = 10 μm.

Whole Rock Composition	Wt. %
Quartz	15
K-Feldspar	2
Plagioclase Feldspar	6
Calcite	25
Dolomite	5
Pyrite	13
Layer Silicates	25
Illite-Smectite	38
Illite-Mica	35
Kaolinite	18
Chlorite	9
TOC	7.8

WELL DEVELOPED LAMINATION IN A BLACK SHALE (EXAMPLE II)

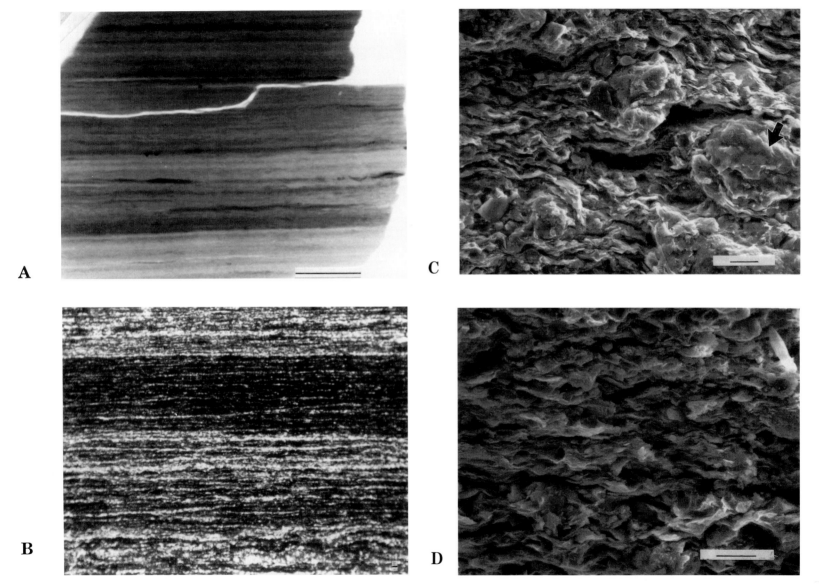

ORGANIC VARIATION IN A SHALE - CLUES TO THE CAUSE OF LAMINATION

JET ROCK SHALE FORMATION (Upper Lias, Jurassic) Ravenscar, Yorkshire, England.

This section presents a detailed analysis of the composition and fabric of a typical lamina couplet of alternating fine-coarse material in the black Jet Rock Shale. The purpose is to show the usefulness of microanalysis in obtaining evidence of the origin of black shale and, in particular the origin of lamination.

Sedimentary Environment: marine, anaerobic

Sedimentary Features: variation in organic content in layers of highly laminated shale.

Geology: Lamination in shale may be due to differences in composition, texture, and color. Rhythmic repetition is due to variations in transport or production of material resulting from fluctuations in currents or flow regions plus tidal and seasonal fluctuations (Reineck and Singh, 1980), turbidity currents (Stow and Bowen, 1980; Piper, 1972), seasonal phytoplanktonic blooms (Gallois, 1976; Hallam, 1975), and interrupted growth on the sediment substrate of microbial mats (Riegel et al., 1986). Wavy lamination described for the Bituminous Shale suggests "stromatolitic" bedding (microbial mats?). The finely- to thickly-laminated character of the Jet Rock Shale suggests another mechanism. During the deposition of the Jet Rock shale, conditions were such that continuous sedimentation of organic-rich clay was periodically interrupted by deposition of silt. Lamination in this example provides evidence of systematic alternation of sedimentary conditions, possibly indicating periodic pulses of coarser sediment deposited rapidly into the anaerobic environment.

Whole Rock Composition	Wt. %
Quartz	26
Plagioclase Feldspar	6
Calcite	3
Dolomite	6
Pyrite	21
Layer Silicates	38
Illite-Smectite	25
Illite-Mica	47
Kaolinite	13
Chlorite	15
TOC (average)	7.8

A.,B. The Jet Rock black shale is fissile in outcrop. One factor influencing this property is the prominent lamination, as seen in x-radiograph A (Scale = 1 cm) and in thin-section B (plane light, Scale = 0.1 mm). In B, the lighter areas are silty, whereas the darker layers consist of clay and organics. Notice the variation in layer thickness. An interesting observation in our study of lamination in black shales is that lamina thickness is not regular but varies vertically throughout a unit.

C_1. SEM photograph of microfabric of a typical silt-rich layer. Notice the abundant fine silt grains (large arrows). Scale = 10 μm.

C_2. SEM photograph of the microfabric of a dark organic-clay rich layer directly overlying the zone shown in C_1. Preferred platy particle orientation dominates. Some clay-size quartz grains are present. Pyrite framboids (small arrows) occur in both C_1 and C_2. Scale = 10 μm.

D. A thin diamond coated dentist's grinding disk was used to carefully extract material from each layer of a light-dark couplet. Thickness of the couplet varies but averages 0.2 mm, therefore, extreme care was taken so as not to mix layers during sampling. The fine powder of each layer was analyzed for TOC by a combustion technique. TOC of the clay-organic layer is 11.0% and 4.1% for the silt layer.

ORGANIC VARIATION IN A SHALE

DESCRIPTION OF A TYPICAL SHALE COUPLET

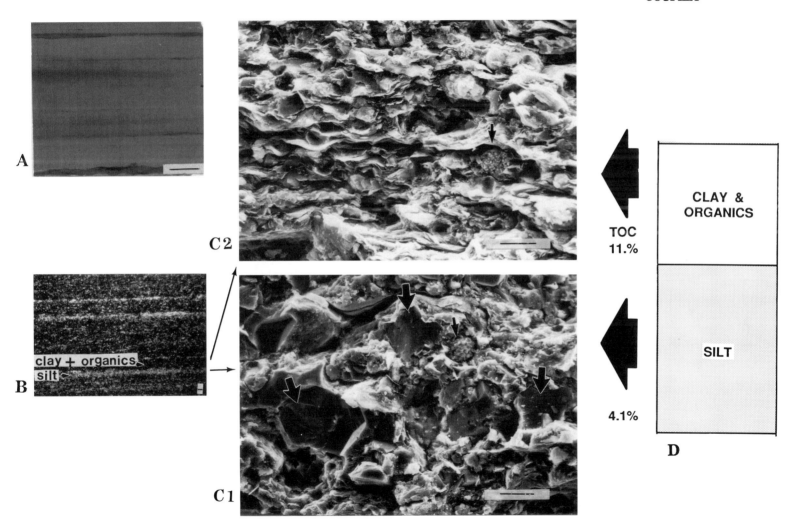

LAMINATED SHALE FROM BOTTOM-FLOWING, LOW DENSITY TURBIDITY CURRENTS

SUNBURY SHALE FORMATION, (Mississippian) Gallia County, Ohio.
This thickly laminated, dark gray shale exhibits fabric evidence of having been deposited from low density, bottom-flowing turbidity currents.
Sedimentary Environment: shallow marine shelf, dysaerobic to aerobic.
Significant Features: particle size variation in adjacent laminae; evidence of erosion by bottom flowing currents.
Geology: The Sunbury Shale was deposited in the Ohio Basin during progradation of the Berea Delta. Lamination resulted from deposition from episodic pulses of low density, low velocity turbidity currents. The fine-grained material was deposited at the distal margins of the flow. Piper (1978) has described how laminated silts and muds may form from turbidity currents. He indicated alternating laminae represent deposition of bedload silt followed by deposition of suspended mud. The Sunbury Shale may have formed at the distal margin of low density turbidity currents during periodic alternating influxes of silty and clayey sediment. The coarser sediment is associated with the turbidity current episode whereas the fine-grained units represent interturbidite or hemipelagic sedimentation.

Whole Rock Composition	Wt %	
Quartz	27	
K-Feldspar	<1	
Plagioclase Feldspar	3	
Siderite	2	
Pyrite	5	
Layer Silicates	58	
Illite-Smectite		30
Illite-Mica		57
Kaolinite		4
Chlorite		9
TOC (whole rock) =	4.5	
TOC Zone 3 =	1.2	
TOC Zone 2 =	0.1	
TOC Zone 1 =	0.3	

A. This X-radiograph displays alternating silty and silt-poor layers. TOC content is high enough to give a dark gray color to the rock although TOC varies considerably between individual lamina. Dark spots at the bottom and top of zone 2 are pyrite. Notice the variable thickness of laminae which appear to have parallel contacts in x-radiograph (however, see contact detail in thin-section, C). Scale = 1 cm.

B. SEM micrographs illustrate vertical variations in microfabric within the laminated shale. Preferred platy particle orientation dominates in clay-rich lamina 1 and 3 (B_1 and B_3). Between them is a 2 mm thick silt-dominated pyrite-bearing zone (see B_2). In this zone clay is randomly oriented owing to the higher silt content (arrows at silt grains); however, this randomness may also be produced by rapid deposition of flocculated clay. Micrographs show the fabric and size variations between layers, which cannot be observed readily by x-radiography. Scale = 10 μm.

C. Thin section of the contact between zones 1 and 2. Notice the miniflame structure (arrow) indicating erosion of fine sediment in zone 1 by bottom flowing current which deposited the silt in zone 2. Crossed nicols, Scale = 0.1 mm.

LAMINATED SHALE FROM BOTTOM FLOWING, LOW DENSITY TURBIDITY CURRENTS

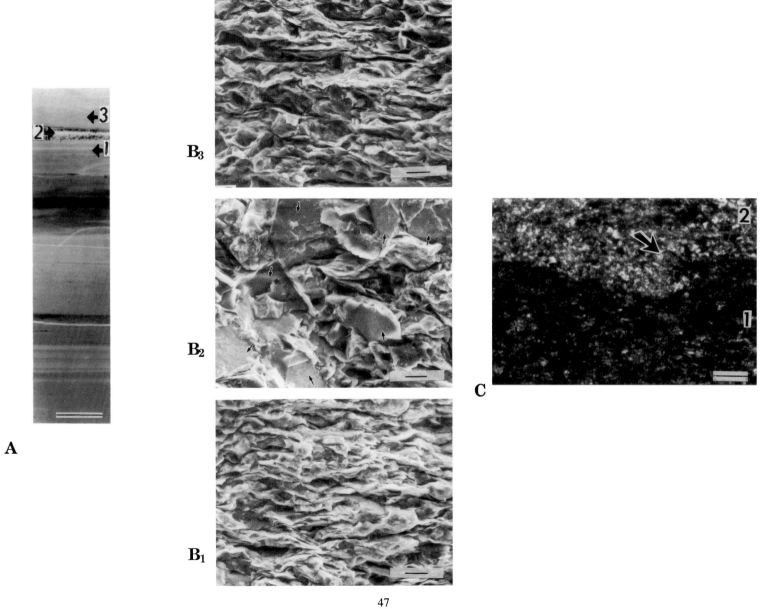

BIOTURBATION

HURON SHALE MEMBER (Ohio Shale Formation, Devonian) Jackson County, West Virginia.

Example shows a sharp contact between bioturbated fabric of a gray shale and overlying non-bioturbated fabric of a black shale.

Sedimentary Environment: marine, dysaerobic changing vertically to anaerobic.
Significant Features: comparison of fabrics of bioturbated and non-bioturbated shales.
Geology: Fabric variations in the Huron Shale show changes in the sedimentary environment from dysaerobic to anaerobic conditions. Random particle orientation in the gray shale is due to bioturbation of bottom sediment. The overlying pyrite zone indicates a geochemical change in bottom conditions which also must have influenced faunal activity. This change is also indicated by the presence of an overlying well laminated black shale with preferred particle orientation. This preferred microfabric indicates burrowing ceased with the development of anaerobic conditions. The change from a dysaerobic to an anaerobic environment appears to have been abrupt, indicated by the sharp gray-black shale contact revealed in the x-radiograph. However, it is difficult to estimate the length of time of this transition from fabric data alone.

Whole Rock Composition	Wt. %
Quartz	30
K-Feldspar	1
Plagioclase Feldspar	3
Siderite	1
Pyrite	4
Layer Silicates	62
Illite-Mica	87
Chlorite	13
TOC	4.1

A. X-radiograph shows well laminated macrofabric (large arrows) of the black shale overlying indistinctly laminated fabric of the gray shale (small arrow). Pyrite grains (dark spots) are concentrated at the interface between the shales. Notice the abrupt contact at the pyritized zone and the difference in macrofabric above and below the contact. Scale = 1 cm.

B. SEM micrograph of well developed preferred orientation of the black shale. A pyrite framboid is indicated by the arrow at the bottom. Scale = 10 μm.

C. SEM micrograph showing randomness of particles produced by bioturbation in the gray shale. Scale = 10 μm.

BIOTURBATION

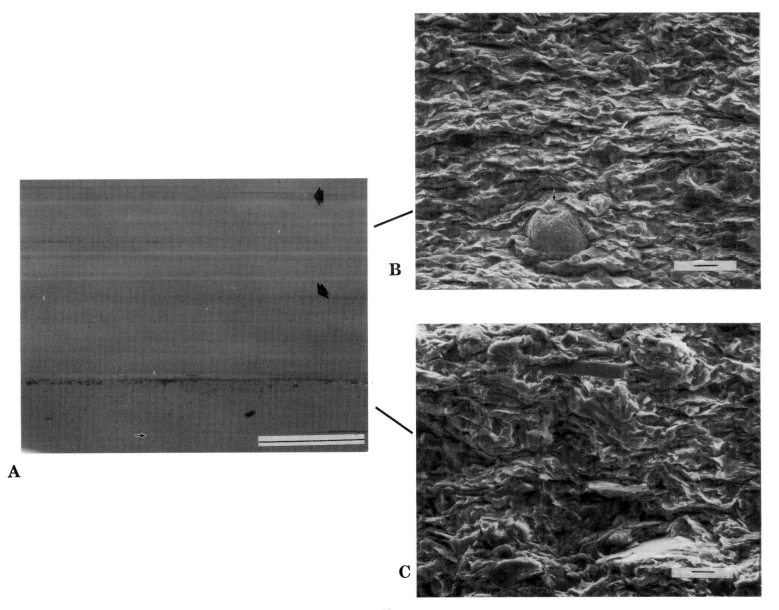

BIOTURBATION - TIERED BURROWING IN SHALE

GRAY SHALE FORMATION (Upper Lias, Jurassic) Port Mulgrave, Yorkshire, England.
   This unit illustrates tiered burrowing and the utility of ichnofabric and sediment fabric analyses in environmental analysis.
Sedimentary Environment: marine, originally anaerobic changing vertically to dysaerobic.
Significant Features: tiered burrowing, preserved remnant shale lamination, bioturbation.
Geology: Trace fossils provide valuable clues to sedimentary conditions that influenced rock fabric. Environmental changes may be evident from juxtaposed suites of trace fossils representing different conditions. Here, for example, *Chrondrites* (which indicates a low oxygen level in sediment) is the only trace fossil in the lower portion of the shale. The depth of penetration of *Thalassinoides* is limited by the anaerobic conditions. It originated in the more oxygenated (dysaerobic?) environment represented in the sequence by an overlying lighter-colored shale. This tiering of burrows indicates progressive change from anaerobic to dysaerobic bottom conditions.

Whole Rock Composition	Wt. %
Quartz	34
K-Feldspar	1
Plagioclase Feldspar	7
Pyrite	10
Layer Silicates	46
Mix Layer (I/S)	16
Illite-Mica	43
Kaolinite	25
Chlorite	16
TOC	2.1

A. Tiered trace fossils in an x-radiograph of the dark gray Gray Shale. Arrow 1 points to a zone of undisturbed shale lamination. Vertically, the lamination becomes progressively more poorly developed. Larger *Thalassinoides* (?) burrows occur at arrow 2. Smaller *Chrondrites* are light gray spots and narrower burrows (examples at arrow 3). Notice that *Chrondrites* burrows penetrate deeply. Scale = 1 cm.

B. Thin section (crossed nicols) of shale with *Chrondrites* burrows. Notice the preservation of remnant lamination in the center of the area. Light areas are burrows and reworked sediment. Scale = 1 mm. Plane light.

C. SEM photograph from area 1 (x-radiograph in A) showing that original preferred orientation is still preserved in non-burrowed areas. Some minor disruption of fabric is apparent. The primary preferred orientation fabric is well displayed at the bottom of the photo. Scale = 10 µm.

D. SEM photograph of a bioturbated portion (zones 2 and 3 in x-radiograph A) of the rock showing fabric within a typical burrow filled with coarser silt grains and randomly oriented clay flakes. Scale = 10- µm.

BIOTURBATION - TIERED BURROWING IN SHALE

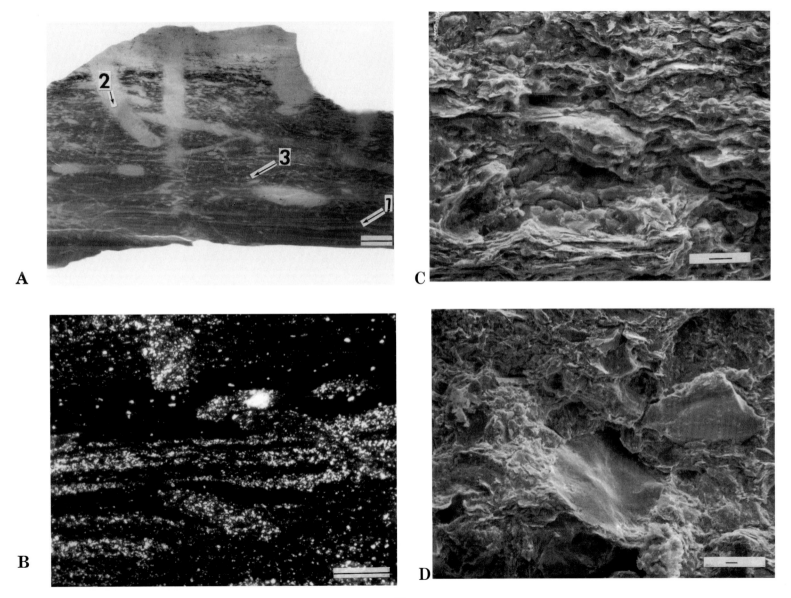

SIGNIFICANCE OF VERTICAL FABRIC VARIATION IN A SHALE

HURON SHALE MEMBER (Ohio Shale Formation, Devonian) Mason County, West Virginia.

This example shows abrupt fabric changes vertically and how they may be interpreted to indicate a temporal change in the sedimentary environment.

Sedimentary Environment: Gray Shale: marine shelf, dysaerobic to aerobic. Black Shale: marine, restricted circulation, anaerobic.

Significant Features: Laminated fabric overlain by unlaminated, bioturbated fabric exhibiting random particle orientation.

Geology: Transgressions and regressions are recognized in Devonian rocks of the eastern United States. A commonly accepted sedimentological model depicts black shale as representing anoxic conditions in deep offshore marine water, whereas gray shale indicates oxygenated, shallower water. This example provides a record of a regression. It is interpreted that black shale deposition occurred in deep water under anaerobic conditions with no infauna present to disturb the original lamination (which is obvious in x-radiograph A) or primary microfabric (shown in the black shale SEM-C). As the sea became more shallow, sufficient oxygen was produced to support some organisms, which mixed the sediment. Bioturbation of the soft, water-laden flocculated sediment produced a homogeneous unlaminated gray shale with random particle orientation well illustrated in SEM-B. Although above the sediment water interface sufficient oxygen was present, anoxic conditions existed in the organic-rich bottom muds. The next unit reconstructs a "moment" in geological history in this Devonian environment.

Whole Rock Composition	Black Shale Wt.%	Gray Shale Wt.%
Quartz	38	39
K-Feldspar	Tr.	Tr.
Plagioclase Feldspar	4	3
Calcite	-	2
Dolomite	-	2
Siderite	-	2
Pyrite	6	<1
Layer Silicates	52	52
Illite-Mica	45	90
Kaolinite	10	3
Chlorite	45	8
TOC	10.5	4.1

A. X-radiograph shows a pyrite-rich zone (large arrow) separating well-laminated, black shale from overlying non-laminated gray shale (small arrow). Notice the burrow at the base of the gray shale (small arrow). Scale = 1 cm.

B. SEM micrograph of random fabric of the non-laminated gray shale. This randomness is suggestive of bioturbation, especially considering the presence of a burrow shown in x-radiograph A. Scale = 10 µm.

C. SEM micrograph of well developed preferred orientation fabric of black shale. Scale = 10 µm.

Samples from the West Virginia Geological and Economic Survey.

GRAY
SHALE
(B)

PYRITE

BLACK
SHALE
(C)

A

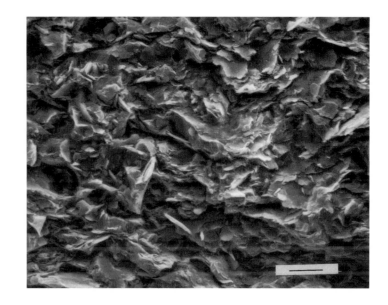

B

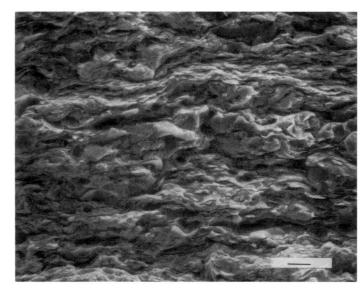

C

A JOURNEY TO "ANOXIA" - RECONSTRUCTION OF AN EVENT ON THE DEVONIAN SEA FLOOR

HURON SHALE MEMBER (Ohio Shale Formation, Devonian)
Mason County, West Virginia.

This section provides more details concerning the usefulness of combining shale macrofabric and ichnofabric (biogenic sedimentary structures) analysis in reconstructing ancient sedimentary environments. Below (left) is an enlarged x-radiograph (Scale = 1 cm) of the same sample pictured in the previous section. Note the shape of the burrow (arrow). The shape offers a clue to sedimentary conditions. Diagrams show, in a sequence of frames, a possible interpretation of the significance of the burrow in the x-radiograph.

Frame A. Organisms during a normal day of burrowing mined the gray sediment, which was probably deposited in a dysaerobic environment. Burrowing was facilitated by the presence of easily penetrated flocculated clay substrate. Any original lamination was destroyed at this time along with parallelism of clay flakes. Below the gray sediment was black organic mud with its original laminae and preferred orientation. Anaerobic conditions prevailed in this organic-rich zone. H_2S generated in this zone played an important role in contemporaneous pyrite formation (shown by diamond symbol).

Frame B. The x-radiograph shows that an organism tunneled down through the gray mud to produce a burrow approximately 0.5 mm wide. Tunneling deeper, apparently the organism approached more reduced sediment. However, in this case, less than a centimeter below the sediment-water interface highly anoxic conditions existed in the area (shown by the diamonds).

Frame C. Upon penetrating this reduced zone, the burrower was surprised when it encountered the top of the anaerobic sediment rich in H_2S.

Frame D. The organism may have quickly stopped tunneling and retreated vertically, digging another escape burrow in the overlying safer, more oxygenated sediment. Notice the abrupt turn in the lower segment of the burrow. Alternately, the shape of the burrow could have been distorted by later compaction of the sediment (flocculated?). What is clear, however, is that burrowing did not proceed into the anoxic sediment of the black mud because of restricted conditions. This example illustrates that macro- and micro-fabric analysis provides another way to view the details of geological history.

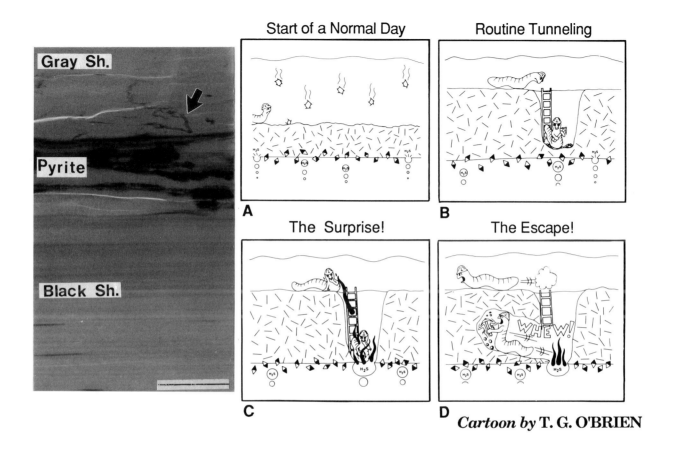

CHAPTER 6
CASE STUDIES OF FABRIC ANALYSIS IN EVALUATING SEDIMENTARY PROCESSES AND ENVIRONMENTS

MARINE REGRESSIONAL FACIES

HUSHPUCKNEY SHALE MEMBER (Swope Formation, Pennsylvanian) Adair County, Iowa.

This example shows changes in shale fabric associated with a marine regression. The Upper Pennsylvanian megacyclothems illustrate the numerous marine transgressions and regressions that occurred in the Midcontinent region of North America. The regressive phase of a cycle in this example is represented by a shoaling-upward sequence of black shale overlain by gray shale, in turn, overlain by marine limestone. Shown here is how fabric changes vertically with environmental changes.

Black Pennsylvanian shales of megacyclothems of the Midcontinent (like the Hushpuckney) are thought to have been deposited in deep water at times of maximum marine inundation (Schutter and Heckel, 1985). Heckel (1977) indicates that the black shale facies, lacking benthic fossils, represents anoxic conditions which developed when vertical water circulation was eliminated by thermocline development as the sea extended over large parts of the shelf during the maximum stand of sea level. Later, as sea level dropped, a typical shoaling-upward sequence formed.

Sedimentary Environment: Transition from deep anaerobic to shelf aerobic marine water.

Significant Features: Lamination, bioturbation, systematic vertical change in fabric.

Geology: The vertical fabric change reflects a change in sedimentary conditions with time. Fabrics along with paleontologic and stratigraphic data support a change from a deeper water sea (black shale) to the more shallow shelf conditions (light gray mudstone). The lamination and parallel particle orientation of the black shale is a good clue that the original primary shale fabric was not altered by post-depositional mixing. Although fabric does not offer a clue to water depth, well developed primary laminae coupled with the high organic content, does indicate anoxic bottom conditions. A change in these conditions with time occurred as evidenced by the gradual vertical modification from preferred to random fabric along with a change in rock color from black to light gray. The highly random light gray portion of the unit provides evidence that the sediment was extensively mixed by infaunal organisms which inhabited the more oxygenated shallow shelf environment. Fabric data is used here as an important supplement in assessing environmental conditions.

Whole Rock Composition	Wt. %
Quartz	40
K-Feldspar	2
Plagioclase Feldspar	4
Calcite	10
Dolomite	15
Pyrite	4
Layer Silicates	25
Ilite-Smectite	42
Illite-Mica	44
Chlorite	14
TOC (most organic-rich part)	20.8

Macrofabric: The Hushpuckney shale in this example is approximately one meter thick interbedded between two limestones. It grades vertically upwards from black well laminated shale to a dark gray shale with poorly developed fissility, then to a light gray bioturbated mudstone.

A. In x-radiograph the black portion of the shale exhibits well developed lamination. The laminae are laterally continuous but vary in thickness. Compositional variation is indicated by tonal differences. Scale = 1 cm. Used with permission SPEM (Society for Sedimentary Geology).

B. This is a polished section of the surface of the light gray mudstone. Fossil fragments are apparent. Notice the highly mottled nature indicative of extensive bioturbation. Scale = 1 cm.

Microfabric: Photos C, D and E show vertical changes in the shale sequence.

C. SEM micrograph shows the black shale is mainly composed of "organic hash." Parallelism of particles prevails. Scale = 10 μm. Used with permission SPEM (Society for Sedimentary Geology).

D. This SEM micrograph of the dark gray shale is more "clayey-organic" in appearance. The sample occurs at the top of the black shale in the transition zone with the overlying bioturbated light gray shale. Silt (arrow) is also present. A gross parallelism of platy material exists indicating primary fabric has been only slightly disrupted. Scale = 10 μm.

E. Randomness of fabric is seen in this SEM micrograph of the light gray upper part of the Hushpuckney unit. Compare this bioturbated fabric with that shown in earlier sections on Bioturbation. Macrofabric data from the polished section also aid in interpreting bioturbation in this sample.

MARINE REGRESSIONAL FACIES

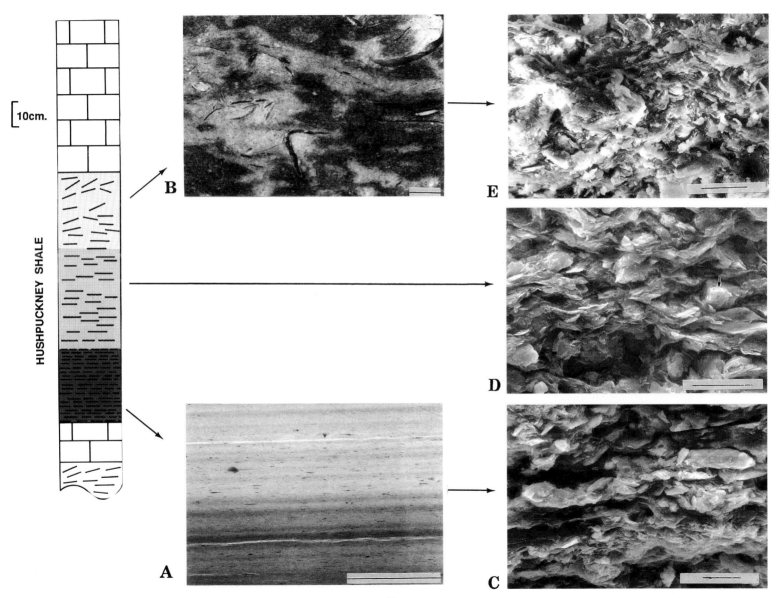

MARINE TRANSGRESSIONAL FACIES

RHINESTREET MEMBER (West Falls Formation) and CASHAQUA MEMBER (Sonyea Formation) (Devonian), Wyoming County, New York.

Fabric variations in shales also record a transgressive event of the Devonian sea in New York State. In this example, the fabrics associated with basin, slope and shelf facies illustrate progressively changing environmental conditions from anaerobic to dysaerobic to aerobic. During Late Devonian time there were numerous phases of deposition with deltaic (Catskill Delta) progradation alternating with transgression of the Devonian sea. During transgression, anaerobic conditions (represented by black shale) shifted laterally toward the east (shore) so that the aerobic shelf became first dysaerobic, then anaerobic. Byers (1977) traced the Middlesex Shale unit (Devonian) laterally from western to eastern New York and found lithologic changes which support this interpretation - "...the main basin remained in an anaerobic state, even while aerobic conditions prevailed on the shelf..." A marine transgression is one explanation for the vertical lithologic changes in the Upper Devonian. Shown here are the shale fabrics which reflect this transgression.

The Cashaqua Shale grades vertically from a poorly fissile, green shale to a moderately fissile gray shale in the Wyoming County, New York Section. It has a dominant pelagic fauna with few infaunal forms, although trace fossils are common. The overlying black Rhinestreet shale is very fissile, finely laminated, and contains few fossils.

Sedimentary Environment: Marine-shelf (aerobic); slope (dysaerobic); basin (anaerobic).

Significant Features: Vertical fabric change associated with organic carbon and shale color change.

Geology: The change in fabric from randomness of particles at the base of the Cashaqua to preferred particle orientation in the overlying Rhinestreet Shale results from a progressive decrease in biogenic activity resulting from a change in sedimentary environment. Bioturbation, prominent in the lower Cashaqua, occurred on the shelf in well-oxygenated water where organisms mixed the sediment and destroyed the primary fabric, producing the fabric observed in figures A and D. Byers (1977) indicated the Devonian sea in the study area was density stratified, producing an oxygen depletion in the deeper part of the basin. As the sea transgressed, the depositional area represented here became more restricted in oxygen content. The gray portion of the Cashaqua represents the marginally oxygenated environment (dysaerobic condition) which developed upon deepening. That bioturbation ceased or was greatly restricted is indicated by the preservation of original shale fabric (see lamination in figure B). Apparently, oxygen levels were still sufficient to allow organic matter to decompose (note the lower TOC values of the Cashaqua shale).

It is interpreted that, with further deepening of the sea, sediment was deposited in an oxygen deficient environment and the black Rhinestreet Shale formed under deeper basin conditions. Under these anoxic conditions organic matter was preserved (e.g. TOC = 3.6%) and biogenic activity ceased, allowing preservation of lamination and preferred particle orientation in the black shale (Figs. C and E). Thus, the petrographic and microfabric details of these two shales reveal important clues in interpreting a marine transgressive sequence.

Whole Rock Composition			
	Wt. %		
	Green Cashaqua	Gray Cashaqua	Rhinestreet
Quartz	33	35	33
K-Feldspar	<1	1	1
Plagioclase Feldspar	4	6	9
Calcite	4	3	1
Pyrite	-	2	3
Layer Silicates	58	54	53
Illite-Mica	77	77	79
Kaolinite	2	3	3
Chlorite	21	20	18
TOC	0.8	0.6	3.6

A. A photomicrograph of a thin-section of the green Cashaqua shale reveals its highly bioturbated nature. Scale = 0.1 mm. Plane light.

B. Well preserved "thick" lamination is common in the gray upper Cashaqua shale. Photomicrograph shows evidence of current lamination with low angle cross-bedding. Scale = 0.1 mm. Plane light.

C. Well developed "fine" lamination prevails in the Rhinestreet black shale. Scale = 0.1 mm. Plane light.

D. SEM microphotograph of random bioturbated fabric typical of the green Cashaqua shale. The arrow points to a silt-size quartz grain. Scale = 10 μm.

E. SEM microphotograph of preferred orientation at the base of the black fissile Rhinestreet shale. Scale = 10 μm.

MARINE TRANSGRESSIONAL FACIES

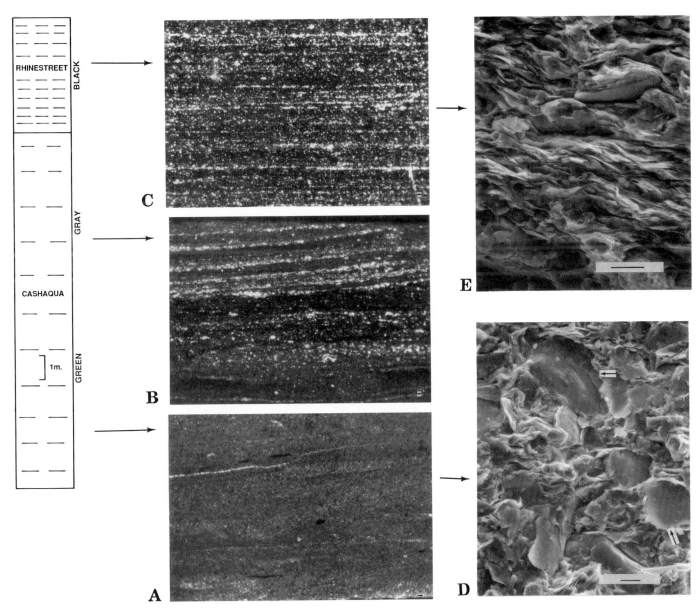

FLOODPLAIN-PALEOSOL FACIES

IVISHAK SANDSTONE (Triassic) Prudhoe Bay Field, North Slope, Alaska.

The Ivishak Sandstone of the Sadlerochit Group is a fluvio-deltaic deposit comprised of sandstones, conglomerates and shales (Atkinson et al., 1988). This sandstone produces oil at Prudhoe Bay Field, the largest oil and gas field in North America, with original reserves in place of over 22 billion barrels of oil and 40 trillion cubic feet of gas. Facies in the field include (1) subaerial, fluvially-dominated coastal plain, (2) transitional, fluvio-marine deltaic, and (3) marine reworked transgressive shelf deposits, each characterized by a different suite of sedimentary textures, structures, and stratification sequences.

Core samples described below are from well DS-3-10 of Atkinson et al. (1988) Floodplain/Pond facies association. The rocks are mainly siltstones which were deposited in low physical energy environments marginal to active stream courses in a floodplain setting. Rocks range in color from dark brown-grey to red-brown owing to variable iron mineral content. They often possess a distinct color mottling attributed to development of soil profiles typical of modern-day floodplains where sedimentation rates are relatively high and groundwater levels fluctuate seasonally. The more mature paleosols exhibit increased levels of mottling and more intense red coloration than immature paleosols.

Sedimentary Environment: Floodplain.

Significant Features: Remnant lamination found in floodplain mudstones, but a lack of lamination in paleosols at the thin-section scale; random fabric found in mudstones and paleosols at the SEM scale.

Geology: Floodplain mudstones exhibit primary laminations at the thin-section scale, which is destroyed during soil formation. At the SEM scale, floodplain mudstones appear to exhibit a primary random fabric, which becomes progressively more random during soil formation. Paleosols are finer-grained, perhaps owing to soil-forming processes.

Whole Rock Composition	A 9320 Wt %	B 9460 Wt %	C 9484 Wt %	D 9875.2 Wt %
Quartz	33	43	41	36
Plagioclase	1	1	1	3
Dolomite	-	1	Tr	-
Siderite	10	1	7	29
Total Silicates	55	55	51	33
Illite-Smectite	53	24	43	22
Illite-Mica	21	46	30	58
Kaolinite	24	26	24	18
Chlorite	2	3	3	2
TOC	0.9	0.6	0.7	.2

A-B. SEM micrographs of the random microfabric in silty mudstone. A = channel abandonment mudstone (sample 9320) and B = unmodified floodplain mudstone (sample 9460). Scale = 10 μm.

C-D. SEM micrographs of very random microfabric of paleosols developed upon the fluvial muds. The paleosols are finer-grained and lack abundant quartz grains compared to the channel and floodplain mudstones. C = immature paleosols (sample 9484.9) and D = mature paleosol (sample 9875.2) Scale = 10 μm.

E-F. Thin-section views showing the contrast between macrofabrics of unaltered mudstone and paleosol. Notice the mottled nature of both samples. E = channel abandonment mudstone (sample 9320). F = immature paleosol (sample 9484). Remnant lamination is observed in the unaltered mudstone (E), whereas soil-forming processes have destroyed this lamination in the paleosol (F). Scale 1 mm. Plane light.

FLOODPLAIN-PALEOSOL

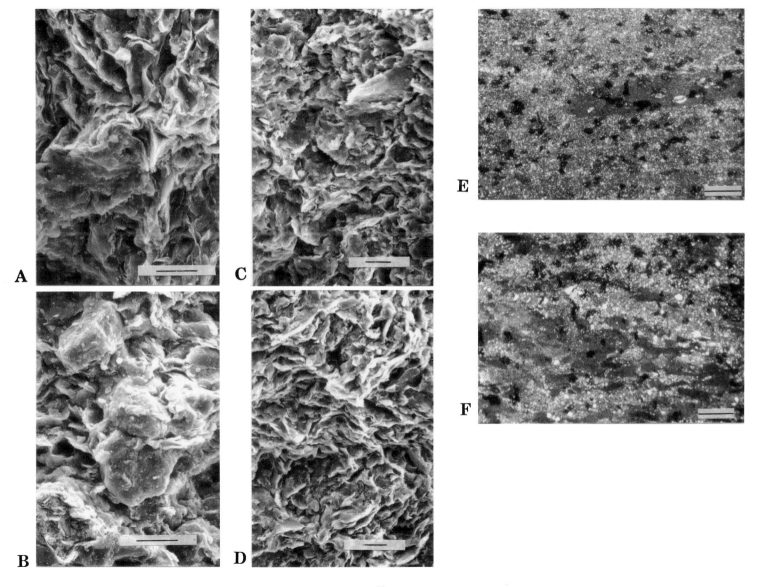

EVAPORITE FACIES

GREAT SALT LAKE (Pleistocene-Holocene) Great Salt Lake, Utah; BRISTOL DRY LAKE (Pliocene-Holocene) San Bernadino, County, California; CLEAR FORK FORMATION (Permian) Swisher County, Texas.

This section displays the common microfabric in mudstones of evaporite facies. The fabric of the Great Salt Lake sample shown is from sediment between 14,500 and 12,700 year B.P. in a saline lake of shallow restrictive circulation (Spenser et al., 1984). The Bristol Dry Lake mudstone was deposited by sheetflood and by suspension settling from ponded floodwater in a closed basin playa setting with ephemeral or perennial shallow saline lakes (Rosen, 1989). Age dating of adjacent tephra indicates the Bristol Dry Lake mudstones were deposited 2-3 million years B.P. The third example, the Clear Fork Formation of the Permian Basin, was deposited in an ancient coastal-sabkha salt pan environment (Handford, 1981). Presley and McGillis (1982) describe portions of the unit as forming on a low relief supratidal or evaporite plain that contained standing water (i.e., a brine pan environment). Horvorka (pers. communication, 1986) emphasizes evaporite deposition in shallow, areally extensive, marine marginal "brine pools" in which mudstones were occasionally deposited.

Sedimentary Environments: Evaporite-continental-sabkha-brine pan; coastal-sabkha salt pan; saline lake.

Significant Feature: Random microfabric of flocculated clay.

Geology: The samples in this evaporite suite all reveal the random particle fabric of primary flocculated clay. Figures A and B illustrate this fabric for Great Salt Lake mudstone. The sediments are conspicuously laminated in hand sample, which indicates a lack of bioturbation and a primary flocculated fabric of the sediment. The random fabric (Fig. C) found in laminated mudstone of the Clear Fork Formation is also a good indicator of primary deposition and preservation of floccules. There is a striking similarity between the flocculated fabric of the Clear Fork and of the Bristol Dry Lake mudstone (Fig. D).

All of the examples illustrate that the typical microfabric of mudstone in an evaporite environment is characterized by an open texture of randomly oriented clay flakes--a fabric produced by the deposition of flocculated clay in saline water. This fabric is contrasted to the preferred particle orientation found in shales from various open-marine environments.

Whole Rock Composition	Great Salt Lake	Wt % Clear Fork Fm	Bristol Dry Lake
Quartz	20	28	24
K-Feldspar	1	4	13
Plagioclase	1	8	
Calcite	15	0.0	6
Dolomite	8	1	
Siderite	-	-	11
Pyrite	-	-	
Halite	19	42	-
Aragonite	10	-	-
Gypsum	2	-	-
Total Layer Silicates	24	17	46
Illite-Smectite	77	26	22
Illite-Mica	17	34	38
Kaolinite	3	-	3
Chlorite	2	10	4
Corrensite	-	30	
Quartz	-	-	2
Gypsum	-	-	31
TOC	2.1	0.21	0.1

A. SEM micrograph showing the random fabric of flocculated Great Salt Lake sediment. Arrow shows typical floccules; Q points to a quartz grain. Scale = 10 μm.

B. SEM micrograph of flocculated Great Salt Lake clay. Scale = 10 μm.

C. SEM micrograph showing the fabric of Clear Fork Formation mudstone. Arrows point to flocculated clay masses. Notice the zones of face-to-face flocculation. Scale = 10 μm.

D. SEM micrograph of Bristol Dry Lake mudstone. Arrow points to floccules. Scale = 10 μm.

EVAPORITE ENVIRONMENT

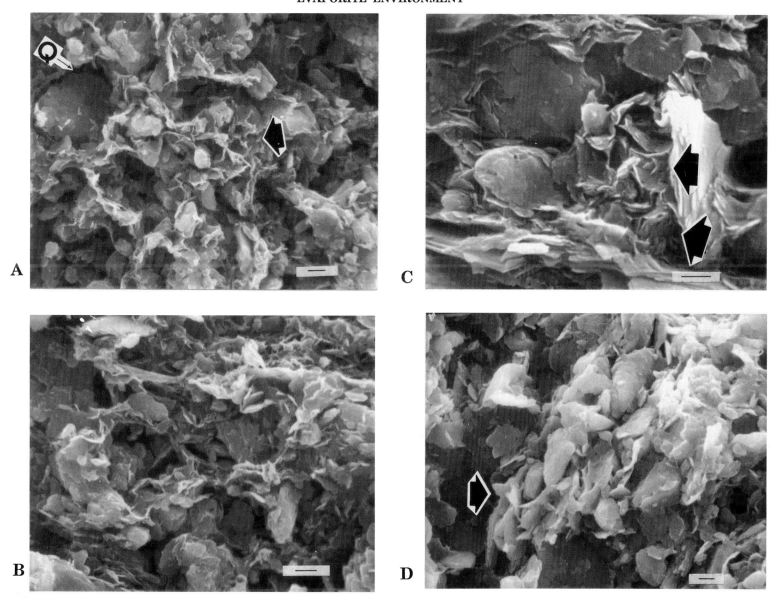

TIDAL FLAT FACIES

RED BED MEMBERS, MOENKOPI FORMATION (Lower Triassic) Clark County, southern Nevada.

The Moenkopi Formation in southern Nevada and on the adjacent Colorado Plateau represents a gradual transition from continental fluvial sedimentation in Arizona through intertidal and shallow marine deposition in western Nevada and Utah (Reif and Slatt, 1979). The red bed members in Clark County are muddy tidal flat sediments deposited in a tectonically stable area of very low relief and arid climate. The Triassic environment is analogous to the upper surface of the modern Colorado River Delta and the Hwang Ho Delta of China. Seven lithofacies comprise the red bed members of the Moenkopi Formation. They range from laminated mudstone, medium- to thick-bedded sandstone, thin- to medium-bedded sandstone, turbated sandstone to mudstone, gypsum, limestone, and dolomite. Detailed fabric analysis of the laminated and turbated mudstone lithofacies are presented in this section.

Sedimentary Environment: Muddy tidal flat.

Significant Features: Well preserved fine laminations, lenticular cross laminae, preferred orientation of particles in clay lamination, random orientation of particles in convolute bed, original clay floccule and evaporite minerals within mudstone.

Geology: The sedimentary structures pictured above in thin section are common to tidal flat deposits and have been described in the literature (Wunderlich, 1967; Thompson, 1968; Reif and Slatt, 1979). PIN-STRIPE TIDAL BEDDING represents high intertidal to shallow subtidal deposition of fine sediment from suspension in the presence of currents and sufficient coarser-grained bedload sediment (silt) to form ripples. With diminishing tidal current intensity, ripple crests become bevelled, forming horizontal laminae. High sedimentation rates inhibit bioturbation. TIDAL BEDDING is produced in a similar manner, but in the transition zone between intertidal and high tidal mud flats where current activity is more intense and coarser-grained sediment is carried in suspension, giving rise to thicker silt laminae. CONVOLUTE BEDDING is produced by post-depositional slumping of unconsolidated mud, as along tidal channel walls. GYPSUM crystals are the product of evaporation of saline water in an arid environment.

At the SEM scale, preferred particle orientation of platy minerals is well developed in most clay laminae and is significant in suggesting another mechanism for the development of argillaceous rock fabric different from the mechanism of compaction of flocculated clay due to overburden pressure (discussed in previous sections of this Atlas). It is reasonable to assume that the original clayey sediment was transported in this hypersaline tidal flat environment as floccules. Photograph L demonstrates that some of the original floccules (arrows) escaped deformation and were preserved in a clay lamina (fabric shown in Fig. L is from a tidal bedded mudstone, Fig. D). Photograph M, however, shows the more common preferred orientation found in laminated zones. This fabric is preserved in the well laminated portion of a turbated mudstone (see arrow in Fig. G). We suggest this orientation resulted from shearing and disruption of the clay flocs by current action at the sediment-water interface (see Figure N), however more observations and experiments are needed to confirm this process.

Disruption of fabric in a turbated mudstone is attributed by Reif and Slatt (1979) to churning of sediment by burrowing organisms, desiccation, growth of gypsum crystals which nucleated within the sediment, and/or erosion and redeposition of sediment prior to lithification. The features illustrated in figures G-K are due to soft sediment slumping, evaporite mineral growth, and desiccation in high intertidal to supratidal environments where exposure and evaporation are common (no definite bioturbation fabrics are present in these samples).

Numerous sedimentary processes and post-depositional processes operate in the tidal flat environment and fabric analysis of mudstones provides one means to determine them. For example, some tidal flat mudstones in our example have a characteristic macrofabric of fine lamination with zones of lenticular cross-bedding. Others show modified macrofabrics indicated by convolute lamination and/or disrupted lamination. In addition, microfabric varies with process. Preferred orientation in clay laminae indicates reorientation of particles in the original flocculated sediment by bottom flowing currents. At times these currents mixed or resuspended bottom sediment sufficiently to alter its fabric and produced zones of random particle orientation. The latter orientation also resulted from sediment slumping or desiccation and evaporite mineral growth.

In summary, microfabrics of tidal deposits indicate current action which tends to produce oriented clay fabric in clay-rich tidal sediments, whereas slumping, desiccation, and salt growth reorient the primary fabric and produce a more random particle orientation.

Whole Rock Composition	A Wt.%	D Wt.%	G Wt.%	H Wt.%	J Wt.%
Quartz	72	57	44	56	45
K-Feldspar	12	14	12	16	--
Dolomite	4	--	--	10	27
Calcite	--	11	23	--	--
Gypsum	--	--	--	--	22
Hematite	--	--	2	--	--
Layer Silicates	12	19	18	18	6
Illite-Mica	88	81	85	89	--
Chlorite	12	19	15	11	--
TOC	0.1	0.1	0.1	0.2	0.1

A. Thin-section photomicrograph of pin-stripe tidal bedding of a laminated mudstone showing alternations of light-colored silt laminae and darker clay laminae. Note the cross laminations. Scale = 0.1 mm. Plane light.

B. SEM micrograph showing random microfabric of a typical silt lamina in a mudstone with pin-stripe lamination. Scale = 10 μm.

C. SEM micrograph showing preferred orientation of particles within a typical clay lamina in mudstone possessing pin-stripe lamination. Scale = 10 μm.

D. Thin-section photomicrograph of tidal bedding in mudstone (similar to pin-stripe tidal bedding, but with thicker laminae) showing cross-lamination (see arrow), with light-colored silt laminae and darker colored clay laminae. Note the bipolar lateral thinning of silt laminae. Scale = 0.1 mm. Plane light.

E. SEM micrograph showing preferred orientation of particles in clay laminae in tidal bedded sample. Scale = 10 μm. Plane light.

F. SEM micrograph showing random fabric silt laminae of tidal bedded sample. Scale = 10 mm. Plane light.

G. Thin-section photomicrograph of a turbated mudstone showing disrupted laminations (small arrow) above a well laminated, ripple-bedded zone (large arrow). Scale = 0.1 mm. Plane light.

H. Thin-section photomicrograph of convolute bedding (arrow) above a ripple-bedded siltstone. Scale = 0.1 mm. Plane light.

I. SEM micrograph showing random orientation of particles within the convolute bed. Scale = 10 μm.

J. Thin-section photomicrograph of gypsum crystals (arrows) scattered within a turbated mudstone. Scale = 0.1 mm. Plane light.

K. SEM micrograph of a gypsum crystal (arrow) that has grown within a turbated mudstone. Scale = 100 μm.

L. SEM micrograph of a clay floccule (arrow) within a clay lamina in a tidal bedded sample (see Fig. D). Scale = 1 μm.

M. SEM micrograph of preferred orientation within a clay lamina in a well laminated zone of a turbated mudstone like that shown in Fig. G. Scale = 1μm.

N. A model suggesting a possible process of disruption of clay fabric at sediment-water interface by bottom-flowing currents.

TIDAL FLAT

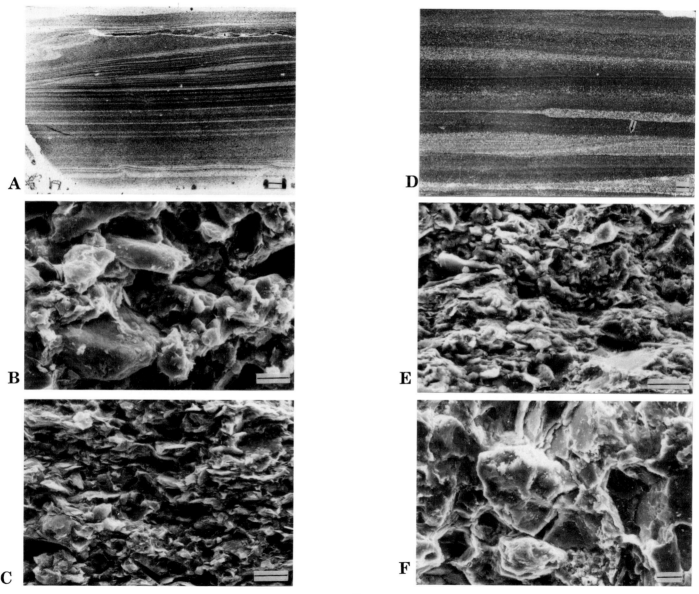

TIDAL FLAT

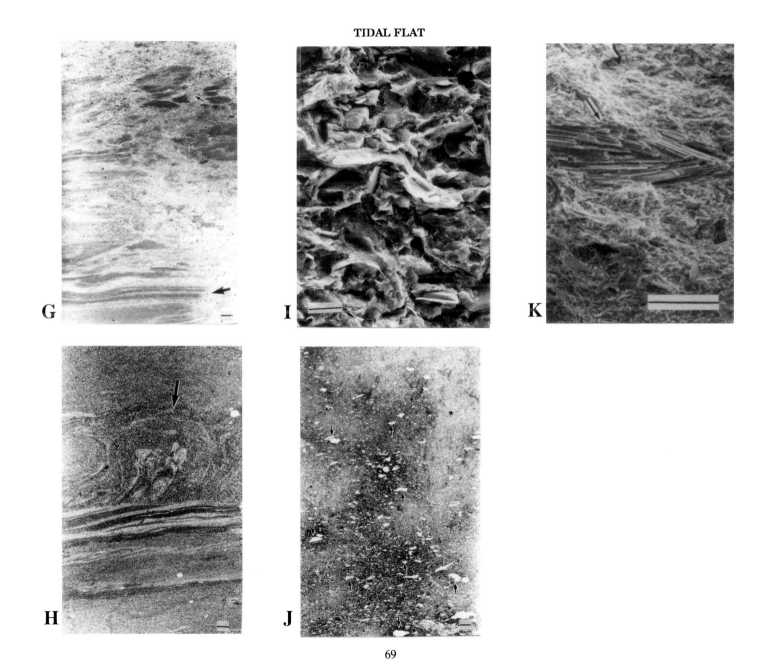

TIDAL FLAT

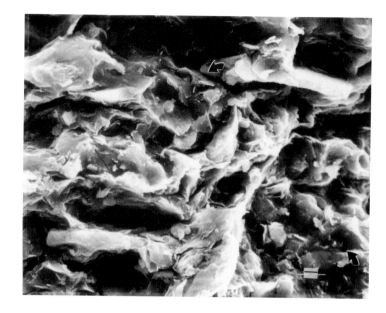

L

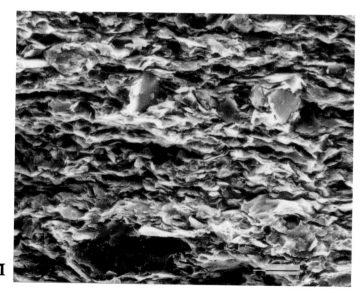

M

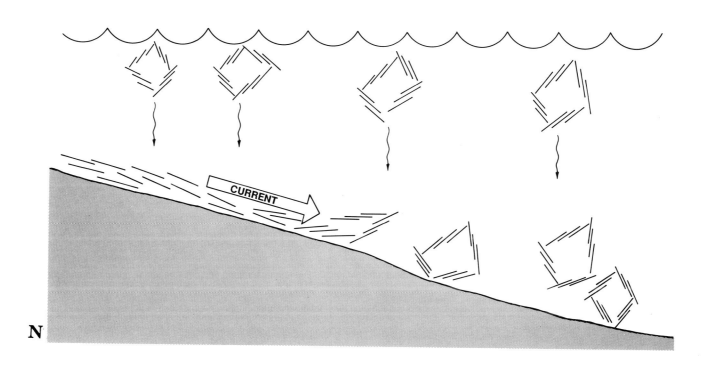

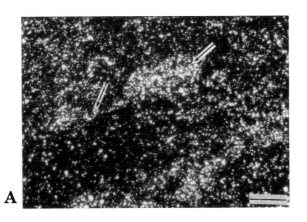

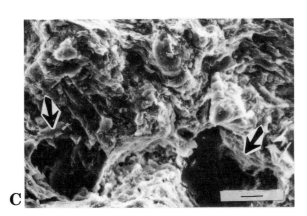

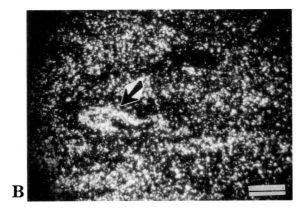

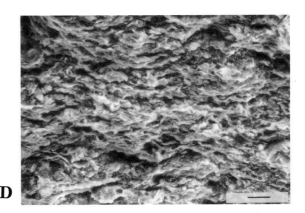

SHALLOW MARINE SHELF

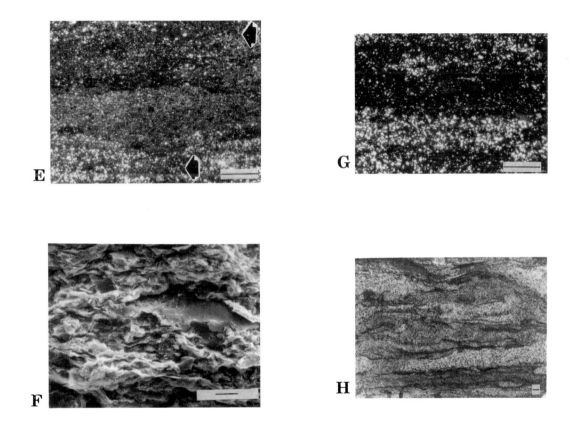

SHALLOW MARINE SHELF FACIES

KUPARUK RIVER FORMATION (Lower Cretaceous) Kuparuk Field, North Slope, Alaska.

A suite of shallow marine shelf fabrics is presented here for the Kuparuk River Formation.

The giant Kuparuk oil field is located on the southern flank of the Barrow Arch, northern Alaska, between the Colville and Prudhoe Bay structural highs (Gaynor and Scheihing, 1988). The reservoir consists of sandstones within the Kuparuk River Formation of Valanginian to Hauterivian age (Lower Cretaceous). The formation is comprised of two distinct depositional sequences separated by a major regional unconformity. Sandstones beneath the unconformity are comprised of several upward-coarsening, lenticular units. Based upon preservation of excellent suites of sedimentary structures in both sandstones and associated shales, they are interpreted to have been deposited during regressive phases of shelf sedimentation by storm-driven along-shelf currents. In contrast, sandstones above the unconformity are comprised of highly bioturbated, glauconitic sandstones and siltstones interpreted to have been deposited on a shelf during marine transgression.

The two samples described below are from a 6.1-m-thick shale interval within the regressive deposits beneath the major unconformity. The shale separates two coarsening-upward sandstone sequences. Foraminifera found in the samples include *Ammobaculites fragmentarious* Cushman, *A. wenonahae* Tappan, *Haplophragmoides topagorukensis* Tappan, and *Trochamimina rainwateri* Cushman and Applin (P. Thompson, 1989, pers. comm.). Although these genera were not abundant in the samples, some inferences can be made concerning depositional environment. The specimens are all quite small, suggesting a stressed environment. Modern analogs of these genera in the Gulf of Mexico are typical of brackish, nearshore environments, possibly tidal marsh or inner shelf.

Sedimentary Environments: Marine, inner shelf.

Significant Features: Lamination and bioturbated fabrics.

Geology: Fabrics associated with marine shelf depositional processes are illustrated in this example. These include current laminations, which give rise to preferred particle orientation at the microfabric scale, and bioturbation, which destroys preferred orientation and gives rise to random microfabric. Marine currents apparently were sufficiently strong to disaggregate original clay floccules as they were being deposited, leading to the preferred orientation seen in photograph B. Evidence of current lamination is seen in the lower portion (small arrow) of Figure D. The random microfabric shown in Figures A, C, and E is evidence of biogenic activity.

Whole Rock Composition	Wt% 6853	6855
Quartz	29	37
K-Feldspar	3	1
Plagioclase	2	2
Siderite	-	1
Pyrite	3	3
Layer Silicates	63	55
Illite-Smectite	32	26
Illite-Mica	28	42
Kaolinite	35	30
Chlorite	6	3
TOC	1.2	1.7

A. SEM micrograph of a bioturbated portion of a silty mudstone showing typical patchy random particle orientation. Scale = 10 μm. Well KRU-1A-13, Sample 6853.

B. SEM micrograph of a silty shale showing preferred particle orientation of a partially bioturbated siltstone. Scale = 10 μm. Well KRU-1A, Sample 6855.

C. Thin-section photomicrograph of the bioturbated sample shown in A. Notice the lack of lamination. Scale = 1 mm. Plane light.

D. Thin-section photomicrograph of the partially bioturbated siltstone sample shown in B. The upper large arrow points to a zone of disrupted, bioturbated lamination, whereas the original undisturbed laminae are preserved in the lower part of the sample (smaller arrow). Scale = 1 mm. Plane light.

E. Thin-section photomicrograph showing a moderately bioturbated zone. Scale = 1 mm. Plane light.

SHALLOW MARINE SHELF

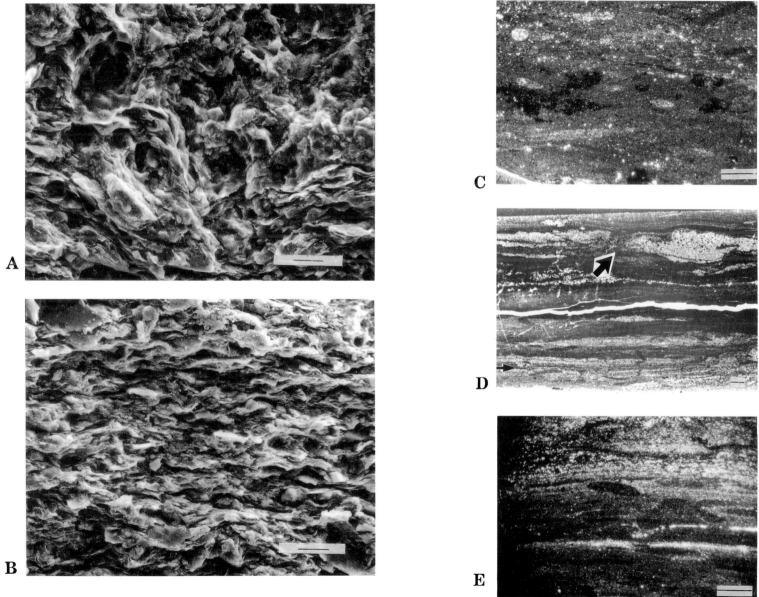

DELTA COMPLEX FACIES

ARGILLACEOUS UNITS OF THE FERRON SANDSTONE MEMBER
(Mancos Formation, Cretaceous) Emery County, Utah.

Macro- and microfabrics (revealed in thin-section photomicrographs and SEM micrographs, respectively) reflect changes of sedimentary environments and processes. This example of clayey units from the Ferron Sandstone (ARCO core 82-6; Thompson, 1985; Thompson et al., 1986) shows the different fabrics of shelf, shoreface, and progradational deltaic wedge facies.

During Jurassic and Cretaceous times, a major north-south trending Western Interior Seaway occupied the interior of the North American Continent. Sediment from the Cordillera to the west was deposited in the basin, forming clastic wedges of fluvio-deltaic, marginal marine, and shelf sedimentary facies. Fabrics of clayey units associated with the Last Chance delta lobe in Castle Valley, Utah (Thompson et al., 1986) are shown in this example. Influencing these fabrics were waves, storms, tides, rivers and burrowing organisms.

Sedimentary environments represented are illustrated in the diagram and labeled A-F.

Sedimentary Environment: Prograding delta complex

Significant Features: Variation in particle size and fabric associated with various sedimentary processes.

Geology: Thompson et al. (1986) indicated that in the Last Chance delta complex, shelf sediments were reworked by currents and bioturbating organisms. Tidal currents and shelf turbidity currents could also have influenced sediment transport. Sample A is from an open-shelf environment. The influence of bioturbation is apparent in the randomness of the fabric. The conspicuous laminations shown in Figures B and C are representative of current lamination as suggested by Thompson et al. for the prodelta shelf environment. The lamination and preferred orientation of the fine-grained sediments in B and C are interpreted as the result of shelf currents prevailing off the mouth of a delta lobe. In this high energy environment of turbid water, sediment mixing would be less due to the restriction on biogenic activity; hence, original lamination fabric could be preserved. Coarser sediments shown in D, E, and F are associated with near shore deposits influenced more by fluvial processes. Sample F, for example, contains relatively large quartz grains in a random matrix of clay which indicates more rapid deposition of riverine sediments on the delta plain (e.g. floodplain or crevasse splay).

Sediment deposited farther away from crevasse splays and washover fans in interdistributary bays exhibit a fabric influenced both by riverine and wave, storm, and tidal processes. Samples D and E maintain a faint lamination (either the remnants of imperfect current lamination and/or due to partially bioturbated sediment) and moderately preferred particle orientation (compare E_2 to B_2); however, the coarser grain size suggests deposition closer to source.

In summary, the fabric and grain size of the argillaceous units in the Ferron sequence vary vertically owing to changing sedimentary processes and environments, which impart recognizable features used in sedimentological interpretations.

References: Thompson (1985), Thompson et al., (1986), O'Brien and Slatt (1988).

Whole Rock Composition:					Wt. %	
	A*	B*	C	D*	E	F
Quartz	72	50	46	50	82	80
K-Feldspar	2	2	2	2	3	3
Plagioclase Feldspar	4	2	2	1	1	1
Calcite	0	7	18	1	0	2
Dolomite	11	20	14	31	0	0
Pyrite	2	2	0	2	2	0
Layer Silicates	11	16	18	13	12	14
Illite-Smectite	14	27	13	9	13	39
Illite-Mica	26	18	22	25	21	18
Kaolinite	49	35	40	49	54	33
Chlorite	11	19	25	16	12	10
TOC	1.5	1.5	2.8	4.4	4.8	0.2

* Average values of numerous samples from these environments.

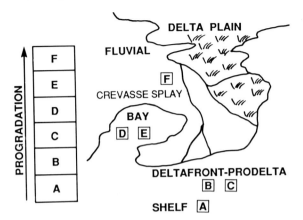

ENVIRONMENTS REPRESENTED IN CORE 82-6

Section Description and Fabric Signatures:

Thin section photomicrographs are shown in column A1 to F1 (A1-E1 = Plane light; F1 = Crossed nicols), Scale = 1 mm. SEM micrographs are in columns A2 to F2; Scale = 10 μm. Note: sample locations with respect to sedimentary environments are indicated in the above diagram and in cross sections (see arrow).

A. (Fig. A-1) - A homogeneous fine-grained texture is exhibited in thin-section. (Fig. A-2) - SEM analysis reveals randomness of platy material, suggestive of bioturbation. Arrow points to quartz grain.

B. (Fig. B-1) - Photomicrograph shows an alternation of dark-(organic-rich?) and light colored (clay and silt) laminae interpreted by their thickness to represent current laminations. (Fig. B-2) - Preferred orientation is well developed in the SEM photograph.

C. (Fig. C-1) - The same fabric as B, however notice the slight disruption of laminae, possibly due to weak currents since mixing by bioturbation is not indicated in the SEM, (Fig. C-2) - which also shows the fine nature and good sorting of the sediment, plus preferred orientation.

D.,E. Samples D and E have high average TOC values (4.4%-4.8%). Organic lenses are shown by dark streaks at E_1 (arrow). SEM micrographs D_2 and E_2 show quartz grains (small arrow) and zones of preferred orientation (large arrow).

F. Both figures show the abundance of large quartz grains (arrows). The small amount of clay present is randomly oriented.

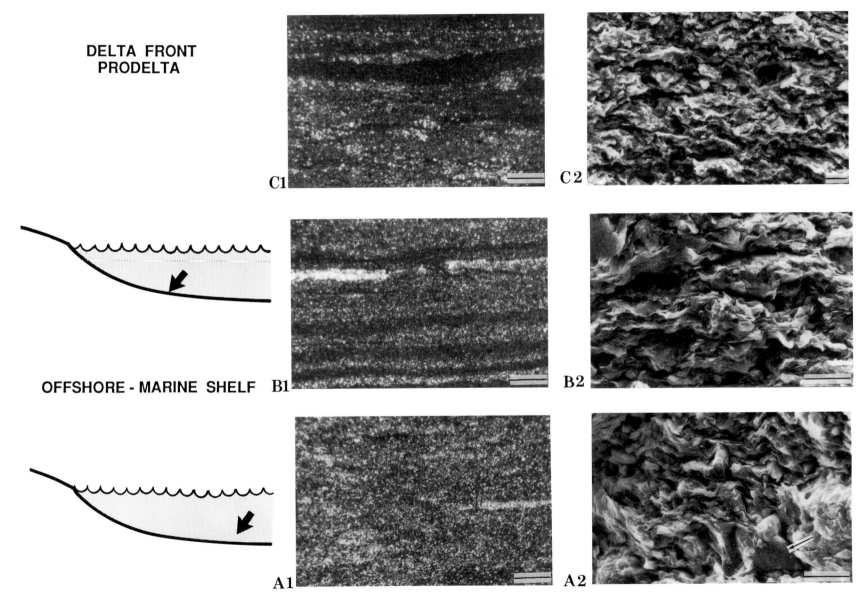

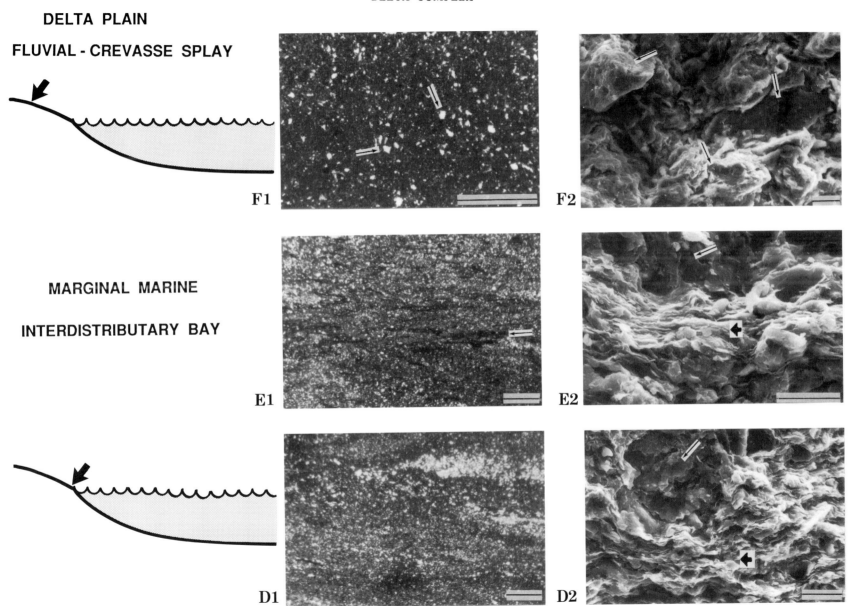

SUBMARINE SLOPE FACIES

MUDSTONE FACIES, COZY DELL FORMATION (Middle Eocene)
Santa Ynez Mountains, southern California.

The Cozy Dell Formation comprises several thousand feet of predominantly mudstone, generally considered to have been deposited in a submarine slope environment. A 120m thick stratigraphic section has been described by Slatt and Thompson (1985) as consisting of a tan, ungraded mudstone facies and a gray, laminated mudstone facies with associated lenticular sandstones. Foraminifera within the mudstones indicate deposition in upper bathyal or slope water depths of 150-500m. The various mudstone fabrics represented here show the influence of processes operating in the marine slope environment. The ungraded mudstone represents an upper slope hemipalagic deposit; whereas the laminated mudstone is an overbank deposit associated with channels on the slope. The depositional setting is analogous to the modern Mississippi River delta front. This settling is shown in Figure E along with interpretations of processes and environments.

Sedimentary Environment: Submarine slope.

Significant Features: Random orientation of particles within the tan, ungraded mudstone; laminae and oriented fabric of the gray laminated mudstone.

Geology: Macrofabrics of the two mudstone facies have been identified by Slatt and Thompson (1985) as comprising units of Stow and Shanmugan's (1980) Mudstone Turbidite Facies. Figure E reconstructs environmental conditions and associated fabrics. The tan, ungraded mudstone (Figs. A, B) is classified as T7 Ungraded Mud in the scheme of Stow and Shanmugan. Its fabric resulted from slow sedimentation either from the distal portion of waning turbidity currents or by hemipelagic sedimentation in relatively quiet water. Its random microfabric (Fig. B) records preservation of primary clay flocs in quiet water of slow suspension settling and burial. The gray laminated mudstone (Figs. C-D) is classified T2-T3 (Stow and Shanmugan) Thin Irregular to Regular Laminae. It resulted from deposition from more proximal muddy turbidity currents, which flowed downslope and was deposited more rapidly (see the diagram on the next page). Preferred orientation and parallel lamination characterize the gray mudstone which formed from slope channel overbank sedimentation. Orientation and lamination probably are a result of a process of shearing of floccules by high energy bottom flowing turbidity currents, in a similar manner to that described earlier for Tidal Deposits of the Moenkopi Formation.

Mineralogy and TOC contents support this interpretation. The higher TOC and pyrite contents of the gray, laminated mudstone are interpreted to be due to higher sedimentation rates and subsequent burial and preservation of organic matter in a reducing environment. In contrast, lower sedimentation rates in a more distal hemipelagic environment would promote oxidation and decomposition of organic matter.

Whole Rock Composition	Tan Mudstone Wt%	Gray Mudstone Wt%
Quartz	31	25
K-Feldspar	6	10
Plagioclase Feldspar	23	24
Calcite	2	3
Pyrite	0	3
Layer Silicates	38	35
Illite-Smectite	0	48
Illite-Mica	43	40
Kaolinite	2	4
Chlorite	3	9
Chlorite-Smectite	52	-
TOC	0.5	0.9

A. Thin section photomicrograph of tan, ungraded mudstone showing the overall massive appearance with the exception of elongate, oriented black organic particles. Scale 10 mm. Plane light.

B. SEM micrograph of tan, ungraded mudstone showing random fabric of sample (Fig. A). Scale = 10 µm.

C. Thin section photomicrograph of gray, laminated mudstone. Notice conspicuous laminations. Scale=10 mm. Plane light.

D. SEM micrograph of gray, laminated mudstone showing parallel orientation of fabric. Small arrows point to quartz grains. Notice parallel clay flakes (large arrow). Scale = 10 µm.

E. Reconstruction of environmental conditions and associated fabrics (A-D).

SUBMARINE SLOPE

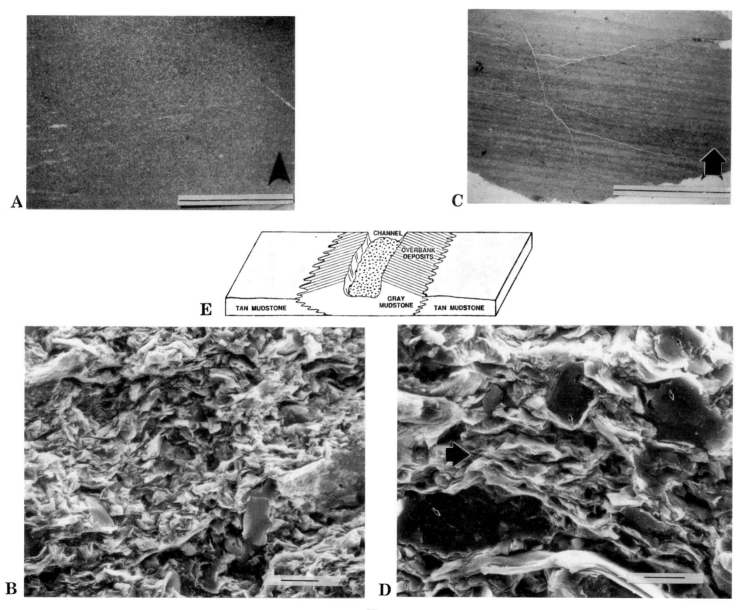

MARINE TURBIDITE FACIES

HURON SHALE MEMBER (Ohio Shale Formation, Devonian) Gallia County, Ohio.

Lamination is a prominent feature of shales and as the previous case study (Cozy Dell Formation) illustrated, may be formed by the turbidity current process. The example of the Huron Shale member provides more evidence of the role turbidity currents played in shale formation. Bottom flowing turbidity currents impart characteristic features to coarser clastics such as graded bedding and various bedforms (e.g. sole markings, convolute lamination, etc.). This unit illustrates that a fine-grained argillaceous rock also may possess these features. Fabric is used here to support a turbidite origin for this shale.

Sedimentary Environment: Marine, slope to basin.
Significant Features: Graded bedding, cut and fill structure
Geology: Fabric evidence suggests that the bottom-flowing turbidity current process could have been responsible for forming the Huron Shale. Evidence of this process is revealed in graded bedding (Fig. B) and "cut and fill" features (Figs. C and D) commonly associated with traction currents. The alternation of coarse silt and finer clayey layers (Fig. A) results in the characteristic lamination of this shale and reflects episodic sedimentation due to periodic influxes of low density turbidity currents onto the slope or flowing out into the basin. The fine-grained nature of the shale also suggests deposition from distal turbidites. Thus, detailed fabric analysis of this shale provides evidence useful in supporting a turbidite origin.

Reference: Schwietering (1979).

Whole Rock Composition	Wt.%
Quartz	36
K-Feldspar	<1
Plagioclase Feldspar	5
Pyrite	5
Layer Silicates	54
Illite-Smectite	14
Illite-Mica	74
Kaolinite	2
Chlorite	10
TOC	1.0 (composite)
	0.8 (silt layer)
	1.7 (dark organic layer)

A. X-radiograph of thickly laminated shale. Scale = 1 cm. Box shows area enlarged in Figure D.

B. Series of SEM photomicrographs showing the vertical microfabric variation in individual laminae: (1) preferred particle orientation of a clay-rich layer; silt is absent; (2) random orientation of clay and silt in overlying silt and pyrite layer. Notice pyrite (black spots) in the upper part of zone 2; (3) the upper part of zone 2 grades into the lower part of zone 3 showing less silt and the return to preferred orientation. Scale = 10 µm.

C. Thin section photomicrograph showing lighter colored silt-rich layers alternating with darker colored organic-rich clay layers. Notice the undulating lower contact at the arrow indicating traction transport and erosion by bottom flowing current. Scale = 0.1 mm. Plane light.

D. SEM enlargement showing the contact between silt layer 2 and the underlying clay layer 1. Notice mini-cut-and-fill feature (arrow). Fig. D is an enlargement of the area shown in the box, Fig. A, and shows details of the contact between layer B1 and B2. This contact suggests deposition of silt from slow moving bottom currents. Scale = 10 µm.

MARINE TURBIDITE

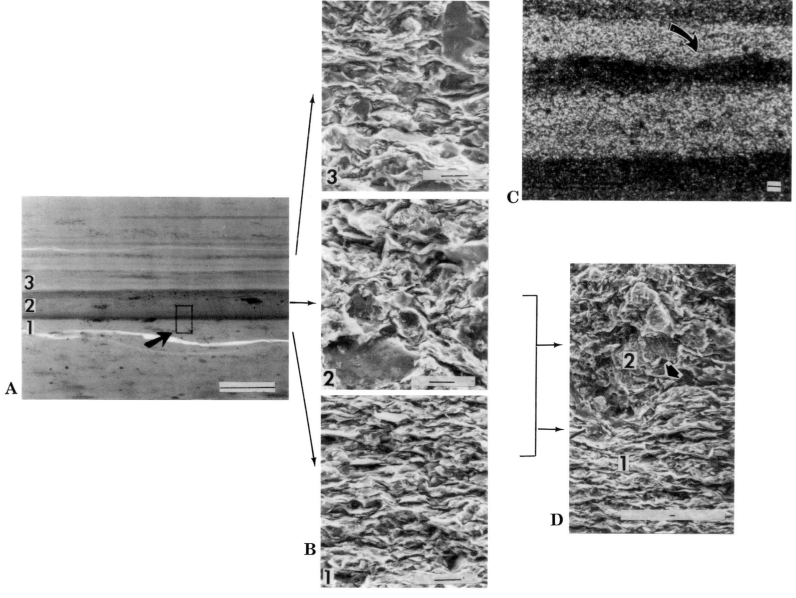

DEEP MARINE TURBIDITE FACIES

PICO FORMATION (Early Pliocene) Long Beach Unit, Wilmington Oil Field, Los Angeles Basin, California.

    The Long Beach Unit of the giant Wilmington oil field consists of hundreds of meters of Miocene-Pliocene age sediments deposited in the northwest-southeast trending, structurally complex and highly petroliferous Los Angeles Basin. The Pico and Puente formations, which occur near the top of the reservoir interval, were deposited in lower middle to upper middle bathyal water depths (500-2000 m) as a series of sediment gravity-flow deposits (Slatt et al., 1988). Thick stratigraphic intervals of mudstone separate intervals of thick-bedded to thin-bedded turbidite sands. Shown here are the macrofabric and microfabric characteristics, representative of deep marine turbidite mudstone (sample from 1068m, well B756I).

Sedimentary Environment: Deep marine turbidite (bathyal depth).
Significant Features: Random macrofabric and microfabric.
Geology: The obvious randomness of fabric shown in the photographs is characteristic of silty turbidite mudstones. O'Brien et al. (1980) found a random clay flake orientation in Pliocene and Holocene turbiditic siltstones from the Boso Peninsula, Japan, indicating that the clay was flocculated, probably due to its high concentration in the turbidity flow. This well-indurated, dark, greenish black mudstone contains few burrows or trace fossils, so the fabrics observed are associated with primary turbidite depositional processes. Burial and compaction of the flocculated clay apparently has not reoriented the particles.

    Note the similarity of this turbidite mudstone fabric to that of the Devonian Huron Shale example. Both show numerous silt-size grains in a random clay matrix. The highest degree of randomness in both samples is found in siltier portions. Our observation is that randomness of fabric increases with silt content.

Whole Rock Composition	Wt %
Quartz	17
K-Feldspar	4
Plagioclase	12
Calcite	5
Pyrite	3
Clinoptilolite	1
Layer Silicates	59
Illite-Smectite	40
Illite-Mica	50
Kaolinite	11
Chlorite	-
TOC	5.7

A. This SEM stereo-pair shows the randomness of the mudstone microfabric. The large arrow points to a quartz grain in the 3-D viewing area. Platy minerals (dominantly clay) are oriented at various angles in this random fabric (small arrows). Notice clumps of clays occur in packets of face-face oriented platelets. Scale = 1 μm.

B. SEM micrograph illustrating randomness of platy minerals intermixed with large silt-size quartz grains (large arrow). Scale = 1 μm.

C. Thin-section photomicrograph of the sample in A-B showing a lack of lamination and poor sorting of turbidite mudstone. Arrows show silt- and sand-size quartz grains. The darker matrix is dominantly clay. Scale = 1 mm. Plane light.

DEEP MARINE TURBIDITE

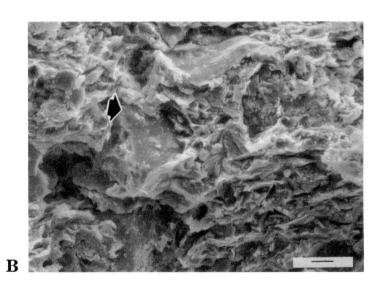

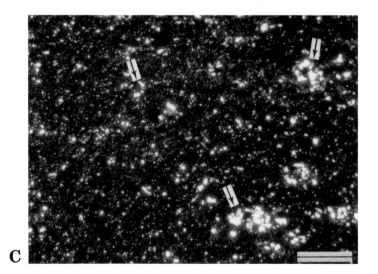

MARINE BASINAL FACIES

GENESEO SHALE MEMBER (Genesee Formation, Devonian)
Ontario County, New York.

In two previous case studies (TIDAL FLAT-Moenkopi Fm. and SUBMARINE SLOPE-Cozy Dell Fm.), mechanisms were discussed which are responsible for formation of lamination in shale. Another example is presented here which emphasizes the significance of the turbidity current process in shale formation. The laminated Geneseo shale of the Appalachian Basin represents a marine basinal facies associated with the Late Devonian Catskill Delta Complex. Ettensohn (1985) indicated that organic-rich facies formed in density-stratified marine waters which produced anoxic bottom conditions. Additionally, evidence of turbidity current deposition is presented by Woodrow (1985). The fabric of the Geneseo shale is described below and a possible mechanism for the origin of the fabric is illustrated in Figure D.

Sedimentary Environment: Marine, anaerobic.
Significant Features: Well preserved fine lamination.
Geology: One characteristic of this finely laminated Devonian black shale is that the silt laminae, composed of one to a few quartz grains, are aligned in parallel layers (Figure A). The parallel character of laminae boundaries, plus the fine-grained nature of the rock, supports deposition in a quiet environment, possibly far offshore and an absence of bottom flowing currents. It is proposed that the alternating layers could be produced by a detached turbid layer mechanism similar to that described by Drake (1971), Pierce (1976), and Stanley (1983). This mechanism is illustrated in Figure D. Ample evidence of coarser turbidites is found in clastic units elsewhere in the Devonian section of New York. In this model dilute sediment gravity flows introduce fine-grained sediment into a density stratified marine basin. The period of each pulse of sediment is unknown (seasonal?). The low density turbid layer flows over the pycnocline because of a density difference of the water (A in Figure D). The mass of fine suspended sediment in the low density turbid layer is unable to overcome the density difference imposed by the pycnocline and flows laterally basinward. At the most distal margins of the flow only the finest silt and clay rain down from an already very dilute detached turbid layer (B in Figure D). Previous to this episodic silty sedimentation, ongoing hemipelagic sedimentation of clay and organics has resulted in an organic-rich bottom sediment layer (B in Figure D). After silt deposition, hemipelagic sedimentation still continues under the prevailing anaerobic conditions (C in Figure D). A couplet of silt-clay laminae in the Geneseo black shale thus could represent episodic sedimentation. Background sedimentation of the organic-clay layers represents the time interval between turbidity current deposition and silt represents the turbid layer event. Alternating fine laminae of silt with organic-rich clay which exhibit parallel contacts seem to be supported by this mechanism. Details of this process are described further by O'Brien (1989).

It is clear, however, that the laminae in this organic-rich marine shale reveal alternating depositional processes.

Whole Rock Composition:	Wt. %
Quartz	39
K-Feldspar	<1
Plagioclase Feldspar	5
Calcite	2
Dolomite	4
Siderite	2
Pyrite	5
Layer Silicates	44
Illite-Mica	76
Kaolinite	2
Chlorite	22
TOC	3.7

A. Thin-section photomicrograph showing light-colored silt laminae alternating with darker organic-clay laminae (arrows point to typical fabric of each lamina seen in SEM). Scale = 1.0 mm. Plane light.

B. SEM micrograph of the highly preferred particle orientation in a typical organic clay layer. Scale = 10 μm.

C. SEM micrograph showing typical microfabric of a silt layer (notice quartz grains, large arrow). Scale = 10 μm.

D. Model for producing lamination by the detached turbid layer mechanism.

B, C and D from O'Brien, 1989. With permission of publisher, Northeastern Geology.

MARINE BASIN

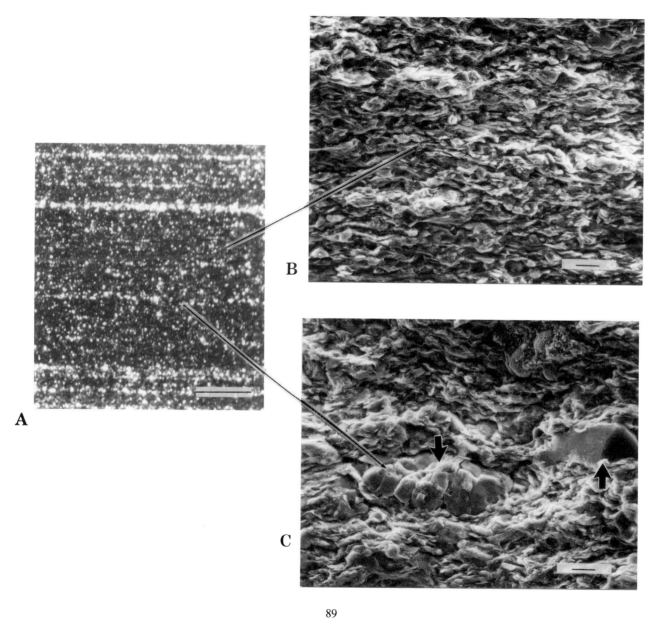

MARINE BASIN

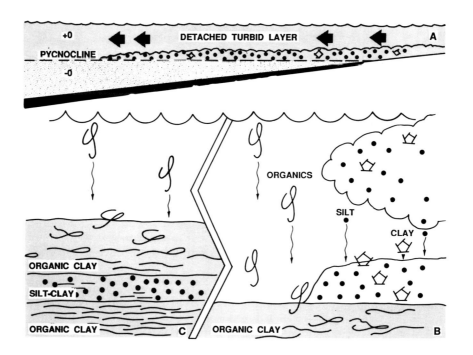

LAMINATION IN A DEVONIAN SHALE PRODUCED BY THE DETACHED TURBID LAYER MECHANISM

CHAPTER 7
FORMATION OF SHALE BY COMPACTION OF FLOCCULATED CLAY--A MODEL

FORMATION OF SHALE FABRIC BY COMPACTION OF FLOCCULATED CLAY

Compaction is a major physical diagenetic process responsible for the final fabric in many argillaceous rocks. This chapter discusses the role of compaction (mechanical rearrangement of particles) in shale formation. It is commonly agreed that most clay-rich sediment settles out in sea water as aggregates or floccules. The "flocculated model" (see section on PREFERRED PARTICLE ORIENTATION IN SHALE) shows how parallel particles result from an original random orientation. Here, a model is described by which the flocculated fabric of sediment may change into that of shale. Presented is a composite of SEM micrographs of clays of various orientations assembled to illustrate hypothetical successive stages of shale formation. Although from different clays from different environments, the examples are useful in providing an idea of particle orientation at different stages of compaction. An actual case study of fabric change with depth in marine sediment is presented by Bennett et al. (1979).

It is our contention that shale fabric may form soon after deposition and during burial as platy particles rotate to a parallel or subparallel orientation. In previous sections we described the role played by bottom flowing currents in shearing floccules and producing particle reorientation. Here we stress the role of compaction due to overburden pressure. Rotation of particles due to overburden pressure is facilitated by the expulsion of pore water which acts as a lubricant for particle movement. This is the key in the shale-forming process. Porosity is reduced during compaction as silt and clay particles assume a denser arrangement. Most pore water is removed early in the compaction history of clayey sediment (Burst, 1967). In fact, Powers (1967) found a decrease in natural water content (expressed in dry weight of sediment) from 115% to 70% from 0 to 3m depth of York estuary muddy sediment. A significant study by Burst (1967) involved the diagenesis of Gulf Coast clayey sediment and its possible relation to petroleum migration. He described three stages of "dehydration" of clayey sediments. Stage 1 is most relevant to our analysis of shale fabric formation. At this stage, pore water and excessive interlayer water are removed by overburden pressure; water expulsion is greatest during this early burial phase. Burst indicated that the original 70-80% water by volume is reduced to 30% in the first few thousand meters of burial of Mississippi delta muds. Hinch (1980) also discussed the rapid decrease in porosity of newly deposited sediment, which he stated may contain 80% or 90% water. He indicated also that in Gulf Coast Tertiary sediments the porosity has been reduced to 25-30% at 800 to 1000m in depth.

Our studies indicate that most shale fabric is a result of early burial. The following SEM photographs illustrate a model of how shale fabric forms.

A. In the diagram, stage A and the accompanying SEM micrograph A1 represent the typical cardhouse structure of flocculated clay--that state in which clayey sediment would be found suspended in sea water. Notice the cardhouse structure is characterized by random domains containing face-face and edge-face oriented flakes and the abundance of voids (dark areas). Flocculated clay has a low-density network of random particles leading to high porosity in this early sedimentation stage.

B. After settling onto the seafloor, the clay cardhouses begin to compact, partially by gravity but mostly due to the overburden pressure exerted by the clay mass itself (stage B in diagram). Those aggregates not ripped apart and resuspended by bottom currents begin the long burial process that will result in shale. The orientation shown in the diagram in stage B and the corresponding SEM photograph (B1) present the possible fabric assumed to be present in the upper meter or so of flocculated clayey sediment. Aggregates at this stage are larger and more densely packed since they are pushed closer together to form a mosaic of random particles. Significantly, particle orientation is still random; however, voids (black areas in B1) are less abundant, indicating partial expulsion of pore water. The clay mass is denser, but is still composed of random particles in stage B..

C. An important change takes place after stage B. Pore water continues to be expelled during compaction, facilitating movement of platy particles to a more stable parallel orientation. The cardhouse fabric begins to collapse. Reorientation proceeds through burial of tens of meters (?) in the upper clay layers. Bennett et al. (1979) observed such fabric changes in Mississippi prodelta sediment and found strong preferred orientation from the original randomly oriented (flocculated?) clay at <120 meters depth. Few studies have been done on the actual mechanism of particle orientation with depth so it is difficult to quantify the exact depth at which the "shale" fabric forms. We suggest that a continuous reorientation of particles occurs from the time of initial burial to some shallow depth. Based upon the studies of Bennett et al., Burst (1967), and Hinch (1980), it seems that "shallow depth" could range from a few meters to several hundreds of meters. The rate and amount of reorientation probably vary depending upon factors such as rate of sediment accumulation (i.e., rate of application of overburden pressure), size and sorting, mineralogy and geochemistry of the sediment. Our model simply suggests that the fabric-forming process is continuous upon burial during which particles reorient as pore water is expelled upon compaction. At stage C (see also SEM micrograph C_1), cardhouses have collapsed and porosity is diminished. Platy material exhibits a parallel to subparallel orientation.

This model is based upon the concept that shale fabric formed from collapsed flocculated clay and stresses the role played by overburden pressure. Another mechanism of particle reorientation (i.e., shearing by bottom flowing current, discussed in earlier chapters) could produce similar results in certain cases. But how does one explain random particle orientation preserved in some non-bioturbated mudstones (such as those discussed in the sections on Evaporite Environment and Marine-Monterey Formation)? The state of our current knowledge is insufficient to unequivocally answer this question, however, there are some possible explanations that are worth mentioning and which may be tested by others. First, if it is assumed that random clay fabric represents that of the original undisturbed flocculated sediment, then it is clear that upon burial and lithification some mechanism was operative which prevented drainage of pore water, because the movement of pore water would have provided a lubricant to facilitate particle rotation. Rapid sedimentation of flocculated clay may have been a factor in preventing effective pore water drainage contemporaneous with sediment accumulation. However, the evaporite deposits discussed earlier also display random fabric and they formed under different conditions which may also provide a clue to the preservation of random fabric. It is known that clays with a given moisture content develop an equilibrium condition which may give them considerable strength (Grim, 1962, p. 240). Thixotropic clay, for example, is weakened when disturbed, but gains strength when undisturbed. It should be expected that the "gel strength" develops in an undisturbed flocculated clay sediment on the sea bottom. This feature also may be observed in the laboratory when sedimented flocculated clays are settled in a beaker and allowed to stand for a period, during which time they become stiffer. With time interparticle bonding (aided by oriented water layers or authigenic cements at contact points) could increase the strength of the clay sediment on the sea floor. If the amount and rate of application of overburden pressure provided by sediment accumulation was minimal during the period the flocculated clay mass developed its maximum strength, then the likelihood of the preservation of randomness of the flocculated mud would increase, thus promoting the development of a mudstone upon lithification. More direct evidence is obviously needed to test these hypotheses.

Our SEM observations do, however, provide a significant clue to an alternate mechanism producing random fabric in mudstones. We found that silt content of a flocculated sediment influences the final rock fabric. Silty shales and mudstones display a high degree of randomness of platy minerals. Because of the size and shape difference between silt and clay grains, the fabric of silty-clayey shale and mudstone is more random than that of a rock composed mainly of clay-size material. It is simply harder for platy clay flakes to assume a parallel orientation in a sediment which also contains larger silt grains.

In silty shale there is also ample evidence which supports a fabric origin by compaction. Figures D and E show that clay cardhouses have been squeezed between silt grains during compaction and have flowed around and between the larger grains. Notice, however, that clay between grains occurs in packets of parallel flakes (i.e., representing the collapsed cardhouses composed of face-face oriented domains). The gross fabric of the total silty-clayey shale sample shown in Figures D and E naturally is not as oriented as the clayey shale shown in Figure C because the silt grains force a more random particle orientation. The fabrics shown in Figures D and E have not formed as individual dispersed clay flakes settled with or on top of silt grains. Even though the samples shown in Figures F and G contain less silt than those shown in Figures D and E, they display that aggregates or floccules of clay, in responding to overburden pressure, were collapsed around and between silt grains. Figures F and G also show how silt grains influence clay orientation by causing the platy particles to "mold" themselves around grains during compaction.

A. SEM micrograph of freeze-dried flocculated illite deposited in a sedimentation tube. Scale = 1 μm.

B. SEM micrograph of Pleistocene clayey sediment from Great Salt Lake, Utah. Scale = 1 μm.

C. SEM micrograph of Jet Rock Shale (Jurassic), Ravenscar, Yorkshire, England. Scale = 1 μm.

D. SEM micrograph of Cleveland Shale Member (Ohio Shale Formation, Devonian), Gallia County, Ohio. Scale = 10 μm.

E. SEM micrograph of Penn Yan Member (Genesee Formation, Devonian), Ontario County, New York. Scale = 10 μm.

F. SEM micrograph of a Tertiary shale from the Gulf Coast area. Scale = 10 μm.

G. SEM micrograph of Huron Shale Member (Ohio Shale Formation, Devonian), Mason County, West Virginia. Scale = 10 μm.

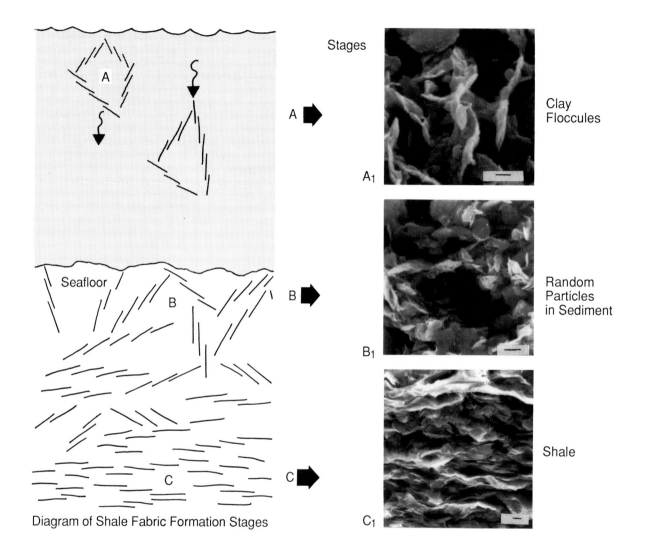

SHALE FABRIC BY COMPACTION (continued)

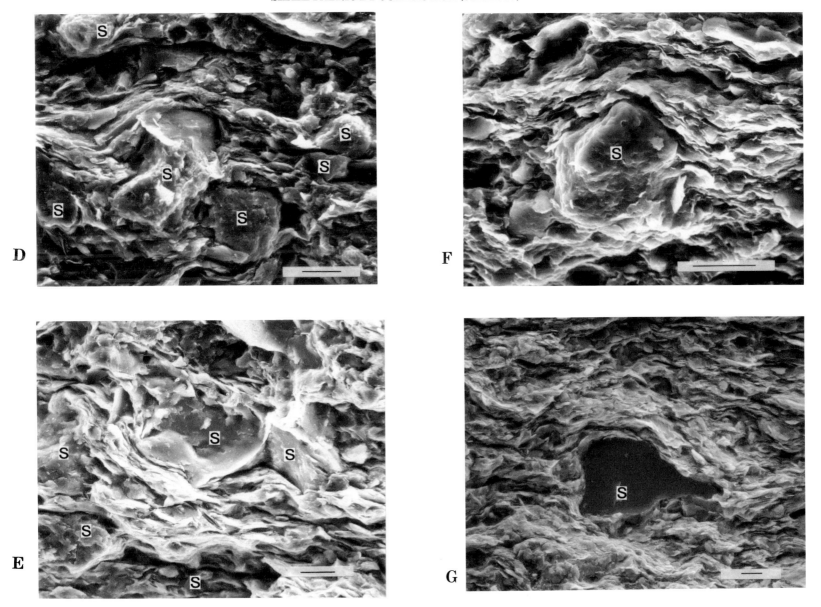

CHAPTER 8
FABRICS OF SOME HYDROCARBON SOURCE ROCKS AND OIL SHALES

INTRODUCTION

This brief summary of petroleum source rocks is mainly summarized from Hunt (1979). A petroleum source rock is defined as a fine-grained siliciclastic or carbonate sediment that in its natural setting has generated and released enough hydrocarbons to form a commercial accumulation of oil or gas. Studies of source rocks have emphasized the organic fraction of the sediments and there is very little published information on the inorganic fraction. Although organic matter is present in most fine-grained sediments and rocks, it is generally considered that at least 0.5-1.5 weight percent organic carbon must be present to form commercial accumulations of hydrocarbons. In general, organic matter content increases with decreasing sediment grain size.

Source rocks are commonly classified on the basis of their chemical and physical composition. The ratio of atomic hydrogen to atomic carbon (H/C) and atomic oxygen to atomic carbon (O/C) can be used to distinguish four major kerogen types in the well-known Van Krevelen diagram (see figure on the next page). Reflected light microscopy of polished rock sections permits recognition of the main constituents (macerals) of the source rock and the association with the host sediment, thus allowing the environment of deposition to be inferred.

Four main groups of kerogens can be distinguished (see figure on the next page). The Liptinite group includes the hydrogen-rich constituents sporinite, cutinite, suberinite, alginite and resinite. These materials are considered to be precursors of oil and gas. Vitrinite constituents include materials derived from the woody and cortical tissues of plants and tend to be relatively hydrogen-deficient and a precursor of gas. The Intertinite group includes a wide range of materials which are hydrogen-poor and considered inert. These macerals can be primary or reworked components and include fusinite, semifusinite, sclerotinite and inertodetrinite. Finally, Amorphinite, or amorphous organic matter, includes a broad range of materials which can have primary and secondary origins. Welte (1974) describes the origin of these materials as random condensation and polymerization of biopolymers, their degradation products and other monomeric materials such as lipids. These materials can have a wide range of chemical composition, ranging from hydrogen-rich to hydrogen-poor, and are considered precursors of oil and gas.

The maturation of organic matter to oil or gas is a function of burial history (time) and geothermal gradient (temperature) in an area, consequently there is not a uniform threshold depth below which hydrocarbons always will be generated. For example, the threshold for intense hydrocarbon generation is about 2,400 m (corresponding to a temperature of 120°C) in the Los Angeles basin and 3,900 m (91°C) in part of the offshore Gulf of Mexico. Typically during burial, source rocks first pass through the "oil generation window," and with increasing burial (temperature and time), gas is generated. For example, in the Eocene of the Gulf Coast, the onset and end of oil generation occur at about 1,500 m and 4,200 m respectively, and the onset and end of gas generation occur at about 2,400 m and 6,900 m, respectively.

Measurement of vitrinite reflectance is a common technique for determining the degree of maturity of hydrocarbon source rocks. The analysis involves the measurement of reflected incident light from the polished surface of vitrinite inclusions in rocks or isolated kerogen using reflected light microscopy and oil immersion objectives. Reflectance values are recorded as mean percent reflectivity or %Ro, measured from a statistically significant number of inclusions. Vitrinite maturation, or reflection, increases with increasing temperature. The lowest level of maturation (Ro) of source rocks associated with generation of oil is about 0.45%, and 0.60% is generally considered as the minimum for generation of commercial oil accumulations. For example, in the offshore Miocene of the Texas Gulf Coast, the onset of major oil generation at about 2,610 m corresponds to Ro of 0.6% and the end of oil generation occurs at about 4,140 m, corresponding to an Ro of 1.35%.

Maturity and reflectance also increase with age. In Louisiana Gulf Coast sediments, the initiation of oil generation (0.6% Ro) is at about 5,490 m (164°C) in Pliocene sediments, 3,540 m (113°C) in Oligocene sediments, and 2,430m (84°C) in Cretaceous sediments.

In this section, examples of some important hydrocarbon source rocks and oil shales are presented to document their fabrics.

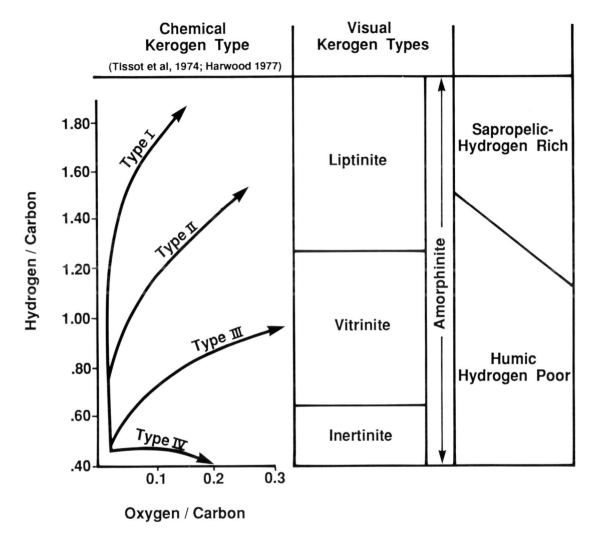

KEROGEN TYPES

MARINE HYDROCARBON SOURCE ROCK

KIMMERIDGE CLAY (Jurassic) Kimmeridge Bay, Dorset, England.

The Kimmeridge Clay is generally accepted as the source rock of North Sea oil. The classic section containing the important organic-rich units outcrops along the Dorset coast of southern England at Kimmeridge Bay (18 km. east of Weymouth). At this section, five alternating lithologies of the Kimmeridge Clay are common: dark gray mudstone (called "clays"), bituminous shale, oil shale, coccolith limestone, and dolomite. This association of lithologies occurs in a cyclic pattern. It is the relationship of the oil shales and coccolith limestones that interests us here. Fabrics of rocks in the Whitestone Band (*Pectinatus* Zone) and the well known Blackstone layer (*Wheatleyensis* Zone) are presented in this case study because they illustrate the fabric of organic-rich shales thought to be hydrocarbon source rocks.

Sedimentary Environment: Marine, anaerobic.

Significant Features: Argillaceous rocks with high TOC and "organic hash" associated with coccolith limestone.

Geology: The extremely high TOC content of the oil shale (e.g. Blackstone layer in this study) of the Kimmeridge Clay, plus the association with coccolith limestones and other alternating lithologies is of significance considering that the Kimmeridge Clay is an important oil source rock. Alternating lithologies indicate conditions of deposition fluctuated periodically. Two major hypotheses are offered to explain the origin of oil shale and its lithologic association with the limestone. Gallois (1976) argues "...that Kimmeridge oil shales (and therefore much of the oil in the North Sea) were formed from algal blooms, in an environment between open ocean and an enclosed marine basin." He states that the oil shales represent algal blooms (probably dinoflagellate) which deoxygenated and poisoned the water producing temporary anaerobic bottom conditions which allowed preservation of organic matter in sediments rich in palynomorphs such as dinoflagellates. The low layer silicate (14%) content of the Blackstone oil shale (indicating minimal sedimentation of land derived clastics) supports this interpretation along with its "organic hash" fabric.

The organic hash is shown in Figure G. Coccolith bands are believed to represent seasonal blooms (Figure D). Coccoliths found in the unit (seen in Figures G and I) were probably swept down to the bottom when the mass of hemipelagic organic material was sedimented.

Alternately, Tyson et al. (1979) indicate that this sequence was produced in a density stratified shallow marine sea. Below the thermocline, bottom water was oxygen-depleted for much of the time; however, there were numerous temporal fluctuations of the interface between a bottom oxygen-depleted layer and the overlying oxygenated water mass (Tyson et al. 1979, refer to the "$O_2:H_2S$ interface"). They also agree that the oil shales formed under predominantly quiet anaerobic conditions, the main difference being the interpretation of the cause of conditions. Seasonal convective currents are believed to have brought nutrients to the surface, stimulating coccolith blooms.

Whatever the exact mechanism, it is clear that anoxic bottom conditions existed periodically during formation of the oil shale and bituminous shale intervals of the Kimmeridge Clay. Most revealing in our analysis is the obvious "organic hash" fabric which characterizes the oil shale, and its high TOC content. These are features commonly associated with anaerobic conditions in a sedimentary environment and are significant in hydrocarbon investigation.

Even though the "organic hash" fabric which characterizes this shale shows a moderately preferred particle orientation, abundant micropores are still apparent (Figure G). It is suggested that this porous network could have been important in facilitating hydrocarbon migration. This observation deserves future study.

Whole Rock Composition	Blackstone Wt. %	Whitestone Wt. %
Quartz	14	27
Plagioclase Feldspar	3	5
Calcite	53	19
Siderite	4	0
Pyrite	12	10
Layer Silicates	14	38
Illite-Smectite	56	45
Illite-Mica	36	33
Kaolinite	6	15
Chlorite	3	8
TOC	46.2	34.5

A. Outcrop photograph of Whitestone Band showing association of bituminous shale (arrow) and laminated coccolith limestone (LS). Scale = cms.

B. X-radiograph of bituminous shale associated with the Whitestone sequence. Scale = 1 mm.

C. Photomicrograph of thin-section of laminated coccolith limestone above shale in Figure A. Light areas, composed of coccolith remains, alternate with darker organic layers, possibly of dinoflagellate origin. Scale = 0.1 mm. Plane light.

D. Coccospheres (arrows) and coccoliths from the Whitestone Band seen in SEM. Scale = 10 µm.

E. Photomicrograph of thin section of a bituminous shale associated with the Whitestone sequence (Figure B). The golden yellow material is interpreted as kerogen of marine algal origin. Clear areas are quartz grains and/or coccolith clusters. The black material is opaque kerogen and/or clay. Notice the high degrees of preferred particle orientation. Scale = 0.1 mm. Plane light.

F. Photomicrograph of a thin section of "oil shale" associated with the Blackstone layer. The identifications are the same as in Figure E. TOC is 46.2% and layer silicate composition is only 14% which makes this rock a "doubtful shale." However, note the parallelism of platy material. Scale = 0.1 mm. Plane light.

G. This SEM micrograph shows the "organic hash" character of the Blackstone "oil shale" in Figure F. The arrow points to a coccolith. Coccoliths occur as a single plate or as clusters (see Figure I). Preferred orientation of platy material is apparent. Scale = 10 µm.

H. SEM micrograph of a spore-like body found compressed in the Blackstone "oil shale." Scale = 10 µm.

I. SEM micrograph of two clusters of coccoliths surrounded by "organic hash" in the Blackstone "oil shale." Scale = 10 µm.

KIMMERIDGE CLAY

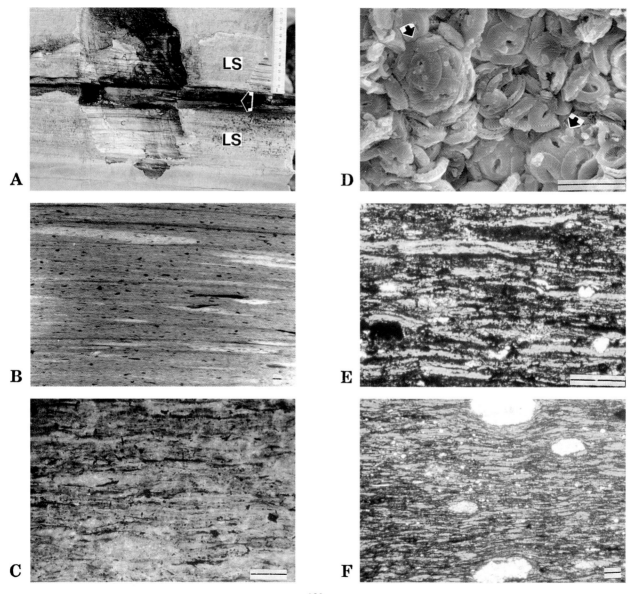

KIMMERIDGE CLAY

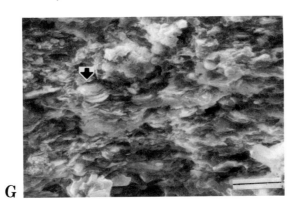

G

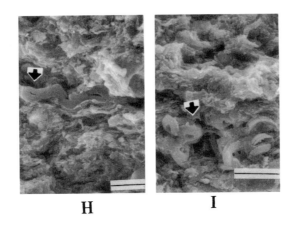

H   I

MARINE HYDROCARBON SOURCE ROCK

WOODFORD FORMATION (Devonian-Mississippian) Carter County and Lincoln County, Oklahoma.

The Woodford Formation is stratigraphically equivalent to organic-rich basinal shales which extend over a large portion of the eastern and central United States. It is a mature, rich petroleum source rock in Oklahoma and western Arkansas. Comer and Hinch (1987) estimate that 27-33% of the total oil generated from the formation has been expelled, amounting to 22 billion bbl of bitumen and 16 billion bbl of saturated hydrocarbons in central and southern Oklahoma. The Woodford Formation is texturally and compositionally variable, owing to temporal and spatial variations in terrestrial and marine organic matter sources and in burial history (Sullivan, 1985). Total organic carbon values are in the range less than 0.1%-26% and degree of compaction and cementation are variable (Comer and Hinch, 1987). All these factors affect the fabric of the shale. Two samples are illustrated in this section, one from Lincoln County (A) and one from Carter County (B), Oklahoma.

Sedimentary Environments: Marine basinal, anoxic.

Significant Features: "Organic hash," moderately preferred microfabric, porous, compressed palynomorphs.

Geology: The Woodford Formation, like the Kimmeridge Clay, is characterized by an "organic hash" fabric which presents a moderately preferred particle orientation. This fabric is not densely packed, however, but is relatively open for a shale. Compressed palynomorphs are common in these samples, as has been noted for the Woodford Shale, in general, by Sullivan (1985). At least a six-fold sediment compaction compressed palynomorphs to their observed shape. Apparently the amount of compaction was insufficient to completely destroy the open network of original floccules. Comer and Hinch (1987) have suggested that hydrocarbon expulsion occurred early in the generative history of the Woodford Formation. The open, porous network observed in these photographs (Figs. A, C, and E) may have promoted migration through the shales and into adjacent carrier or reservoir beds. At a larger scale, lamination zones in the shale (Figures B and C, inset) may have provided an additional avenue of early expulsion of hydrocarbons.

Whole Rock Composition	Wt% A	B
Quartz	68	63
K-Feldspar	4	-
Plagioclase Feldspar	-	3
Calcite	-	10
Dolomite	9	6
Pyrite	7	5
Total Layer Silicates	12	14
Illite-Smectite	44	49
Illite-Mica	47	42
Kaolinite	6	2
Chlorite	3	7
TOC	7.9	5.1

A. SEM micrograph showing "organic hash" microfabric of sample A. Notice that well developed preferred orientation of particles is not obvious although a gross parallelism exists. Notice microporosity. Scale = 10 µm.

B. Parallelism of compressed palynomorphs (small arrows, s) is apparent in this thin-section photograph of sample A. Large clusters of pyrite (small arrows, P) and layers of finely disseminated pyrite (large arrow, P) are common. White circular areas are positions of pyrite dislodged during thin-section preparation. Scale = 10 mm. Plane light.

C. High magnification SEM micrograph of "organic hash" fabric and compressed palynomorph (arrows) in sample A. Scale = 1 µm. Insert is a photomicrograph of the same sample showing a large compressed palynomorph (arrow) in a clay-organic matrix. Scale = 0.1 mm. Plane light.

D. SEM micrograph of a compressed palynomorph in sample B. Scale = 10 µm.

E. SEM micrograph showing a zone of randomness of particles in an otherwise moderately preferred "organic hash" microfabric of sample B. Scale = 10 µm.

WOODFORD SHALE

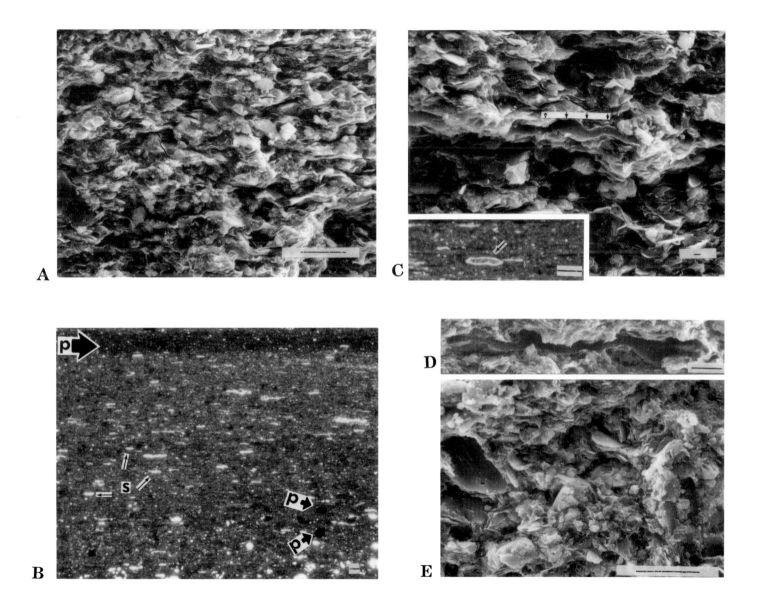

MARINE HYDROCARBON SOURCE ROCK

MONTEREY FORMATION (Phosphatic Facies, Miocene) California.

The Monterey Formation records the deep basinal phase of a major Tertiary cycle of restricted basin sedimentation associated with wrench fault tectonism along the California continental margin (Pisciotto and Garrison, 1981). Fine-grained Monterey rocks are both source and reservoir for major quantities of hydrocarbons both onshore and offshore California. The major facies are calcareous facies of foraminiferal-coccolith shales and mudstones, phosphatic facies of phosphatized foraminiferal shales and mudstones, and siliceous facies of diatomites and diatomaceous mudrocks and their diagenetic equivalents (chert, porcelanite and siliceous mudrock).

The phosphatic facies is illustrated in this section. Phosphate occurs as cryptocrystalline carbonate fluorapatite in irregularly-shaped layers, blebs, and nodules sometimes filling and replacing foraminiferal tests and sometimes without an obvious carbonate precursor (Pisciotto and Garrison, 1981). Rocks of this facies are laminated at the outcrop scale, organic-rich, and average 3-25% silica, 15-50% detrital minerals, 25-75% calcite, 0-20% apatite, and 9-24% organic matter (Isaacs, 1981). According to Pisciotto and Garrison, the high phosphate and organic matter contents, foraminiferal types and abundance, and laminations indicate the phosphatic facies of the Monterey Formation represents hemipelagic deposition near the intersections of a well-developed oxygen minimum zone with the seafloor, probably in outer shelf to upper slope water depths.

Sedimentary Environments: Restricted marine basinal environment.

Significant Features: Random, open microfabric of the phosphatic mudstone, abundant organic, fossil and phosphatic material.

Geology: The macrofabric and microfabric features illustrated here are indicative of suspension settling in a restricted marine environment, as evidenced by the laminations (lack of bioturbation), organic tests, and fluorapatite. Alternating phosphate-rich and relatively less phosphate-rich laminae may represent seasonal alternations in the amount of detrital influx into the basin.

The random, open microfabric of this dark gray organic-rich silty mudstone in SEM micrographs indicates that detrital and organic material were deposited as clumps or aggregates. Some aggregates could, in fact, be interpreted as clay floccules formed by electrochemical attraction which resulted in a typical clay cardhouse (see F). Others could have been aggregates of biogenic and lithogenic material similar to "marine snow" (examples: E and G). Regardless of primary origin, the delicate random fabric of platy materials lacking sharp grain boundaries indicates that physical and chemical diagenesis failed to alter the original sediment fabric. Formed by suspension settling of sediment with a random fabric which was preserved upon burial, the phosphatic facies retained its original porosity--thus probably contributing to primary migration pathways upon hydrocarbon generation. Here again, one sees the association of an open porous network of an argillaceous rock possessing high TOC.

Whole Rock Composition	Wt %
Quartz	22
Plagioclase Feldspar	12
Calcite	23
Pyrite	2
Clinoptilolite	9
Fluorapatite	12
Layer Silicates	20
Illite-Smectite	27
Illite-Mica	62
Chlorite	11
TOC	9.9

A. Thin-section photomicrograph showing lamination in the phosphatic mudstone. Lighter laminae (small arrows) contain a higher phosphatic content than do darker laminae (large arrows). Scale = 1 mm. Plane light.

B. Enlarged area of Figure A. X is in the phosphate-rich zone. Foraminifera tests are abundant. Scale = 0.1 mm. Plane light.

C. Energy dispersive X-ray spectrum (EDX) of area X in Figure B shows major concentrations of Ca, P, and Cl indicating the presence of fluorapatite. Concentrations of Mg, Al, Si, S, and K are associated with clay and iron sulfide minerals. The Ca concentration is partly contributed by foraminifera and coccoliths, and the Si concentration is partly contributed by diatom tests (see Welton, 1984).

D. SEM micrograph showing diatom (small arrow) and foraminifera (large arrow) tests in the mudstone. Scale = 100 μm.

E., F., G., H. This series of SEM micrographs shows the random, open microfabric common in the phosphatic mudstone. E is a characteristic fabric showing abundant random biogenic and lithogenic fragments; silt size grains are also present. The large arrow points to a cluster of coccoliths. Scale = 10 mm. Lithogenic aggregates (F, G, H) display the typical open cardhouse fabric of clay floccules. The edge-face, face-face association is well-displayed in F. Another large aggregate (G) illustrates random particle orientation. A more compressed cardhouse fabric is shown in H. Scale F, G, H = 1 μm.

MONTEREY PHOSPHATIC MUDSTONE

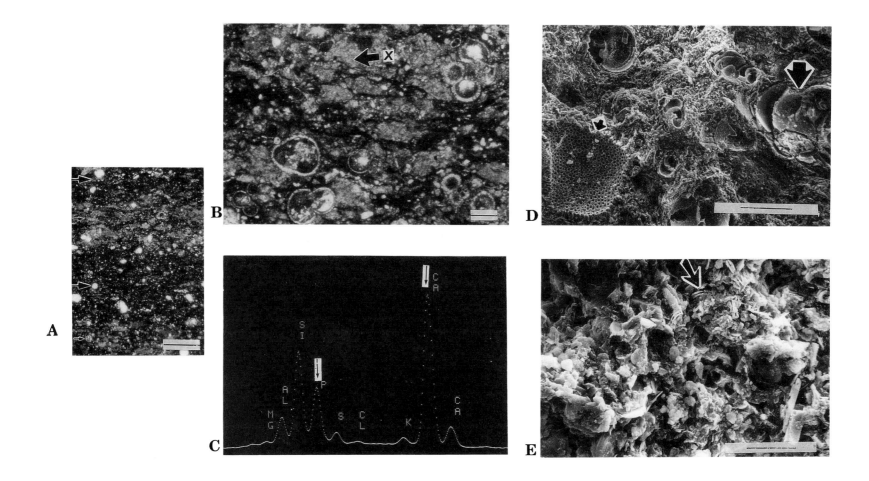

MONTEREY FORMATION

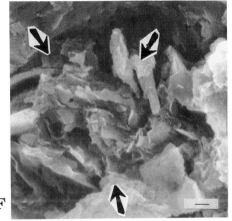

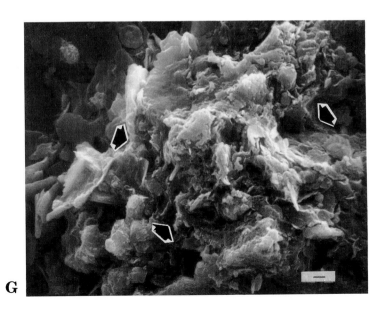

SALINE LACUSTRINE HYDROCARBON SOURCE ROCK

GREEN RIVER FORMATION (Eocene) Parachute Creek, Colorado.

The Green River Formation has a potential shale-oil resource of 8000 billion bbls, spread over an area of 16,500 square miles in the tri-state area of Colorado, Wyoming, and Utah (Shanks et al., 1976; Robinson, 1976). The Green River Formation was deposited from two large Eocene lakes: Lake Uinta in Colorado and Utah, and Lake Gosiute in Wyoming. Post-depositional tectonism has fragmented the lacustrine deposits into the following basins: Uinta, Piceance Creek, Sandwash, Green River, Great Divide, Washakie, and Fossil (Yen and Chilingar, 1976). The sample presented in this study is from the organic-rich Mahogany Zone in the Piceance Creek Basin. Here, the formation is divided, from the base upward, into the Douglas Creek Member, the Garden Gulch Member, the Parachute Creek Member (which includes the Mahogany Zone), and the Evacuation Creek Member. The Parachute Creek Member consists of 130-250 m of gray-black, gray, and brown, organic-rich carbonate rock. Green River oil shales are typically varved, consisting of alternating thin organic-rich and relatively organic-poor laminae. Common minerals include dolomite, calcite, quartz, illite, feldspar, pyrite, analcite, and dawsonite (Robinson, 1976; Shanks et al., 1976). Organic content ranges from 1% to as much as 40% with an average for the Mahogany zone of 16% (Robinson, 1976). The insoluble (kerogen) organic matter consists of structureless, nondescript particles and fragmentary particles of bacteria, algae, spores, pollen, and other organic tissue.

The sediments were deposited in a highly saline lacustrine environment, with shallow-water depths that did not exceed 30 m (Robinson, 1976). The presence of carbonates has been attributed to early diagenesis within a permanently stratified lake. Anoxic conditions prevailed, leading to preservation of the organic debris.

The sample discussed here is from the Parachute Creek member. As the whole rock composition indicates, it really is not a shale, per se, but a carbonate rock (layered silicates <1%). It is included in our discussion because it is a hydrocarbon source rock of wide interest and is commonly referred to as an "oil shale." We will keep this terminology for clarity in our discussion.

Sedimentary Environments: Saline lacustrine, anoxic.

Significant Features: Porous, random microfabric within a laminated dominantly carbonate rock.

Geology: Alternating fine-grained, darker laminae with relatively coarser-grained, lighter colored laminae (Fig. D) are indicative of seasonal or cyclic deposition in a lacustrine environment (Cole and Picard, 1975). Organic content, presence of lamination, and absence of any bioturbation structures indicate that the lake bottom was anoxic at the time of deposition. Robinson (1976) indicated that the organic debris falling into the reducing environment of the hypolimnion was degraded to a gelatinous mass similar to the deposits in present-day lakes described as algal ooze. The open, flocculated-like fabric shown in Figs. A, B, and C may represent remnants of this mass, which settled as organic-inorganic aggregates in a current-free anoxic environment.

Whole Rock Composition	Wt %
Quartz	11
K-Feldspar	4
Plagioclase	9
Calcite	20
Dolomite	45
Pyrite	5
Aragonite	5
Total Layer Silicates	1
Illite-Smectite	68
Illite-Mica	27
Kaolinite	5
TOC	9.9

A. SEM micrograph showing the microfabric of the Parachute Creek member laminated, dolomitic "oil shale." Most apparent is the open, porous fabric comprised of randomly oriented particles. (Note: Compare this fabric to the fabric of the Great Salt Lake clay in another chapter of this volume.) Scale = 10 µm.

B. SEM micrograph at higher magnification showing the porous structure of the "oil shale." The small arrow points to a carbonate rhomb and the large arrow points to typical orientation of platy minerals. Scale = 1 µm.

C. Energy Dispersive X-ray (EDX) analysis of a dolomite rhomb. Arrow points to an aggregate. Scale = 1 µm.

D. Thin-section photomicrograph of the laminated dolomitic "oil-shale." Small arrows point to quartz grains and the large arrow brackets a varve couplet consisting of a lower finer-grained, darker lamina overlain by a coarser-grained (note the greater density of quartz grains), lighter colored lamina. Scale = 1 mm. Plane light.

GREEN RIVER SHALE

FRESH-BRACKISH LACUSTRINE HYDROCARBON SOURCE ROCK

RUNDLE OIL SHALE (Eocene-Oligocene) Queensland, Australia.

The Rundle Oil Shale contains about 600 million bbls of oil in the "Narrows Beds" of the Queensland area (Lindner and Dixon, 1976). The Narrows Beds are confined to a narrow graben about 6 km by 30 km. They are comprised of claystones, mudstones, carbonates, clayey sandstones, lignite and oil shale (Lindner and Dixon, 1976; Hutton et al., 1980). Oil shales are typically well laminated. Organic matter, mainly of algal origin, is present in quantities up to 55%. Minerals include quartz (dominant), feldspar, mica, clay minerals, calcite, siderite, dolomite, and zeolites. Gastropods, pelecypods, fish and reptilian remains, spores, and plants are present in the sediments in addition to an abundance of ostracods, all of which indicate a fresh- to brackish-water lacustrine environment of deposition. The presence of these well preserved organisms, in addition to the uniform and fine-grained, organic-rich lithology, imply continuous, anoxic, quiet water sedimentation. The Rundle Oil Shale exhibits features common to the Green River Formation in that the organic matter in each is almost entirely of algal origin (Hutton et al., 1980). However, the Green River units contain more carbonates and minerals common to a much more saline environment plus the fabric is random. Clay minerals dominate in the Rundle Shale, thus promoting the parallel particle orientation observed in SEM.

Sedimentary Environments: Fresh to brackish lacustrine.

Significant Features: Well-developed preferred orientation of particles.

Geology: This hydrocarbon source rock differs from the others illustrated in this section in exhibiting a well-developed preferred orientation. The open, random orientation of clay particles in the other rocks has been attributed to deposition and preservation of flocculess in marine or saline lake environments. The Rundle Oil Shale was deposited in a quiet, fresh to brackish lacustrine environment. Its organic content (TOC=11.3%) is of algal origin. Hutton et al. (1980) call the Rundle Shale a lamonsite which is a compact laminated organic-rich rock characterized by the presence of Alginite-B (a subdivision of alginite, which is largely of algal origin) as the dominant organic entity. The Alginite B is described by them as finely-banded and lamellar and interbedded with mineral deposits in well laminated order. Their interpretation is that it is derived from microscopic algae, which grew in bloom proportions in shallow warm lakes where they formed a layer of incoherent algal ooze. Our fabric analysis of the SEM seen in Figure A indicates well developed preferred particle orientation which could result either by settling from a very dilute suspension of dispersed clay onto algal mats or by deposition of clay floccules onto the sticky mucilaginous upper surface of bottom algal mats. Continuous growth of algae over each clay or silt sediment increment would result in a laminar fabric composed of alternating layers of organic-rich and clay-silt-rich layers. The result would be a laminated structure similar to stromatolites. A similar fabric is discussed earlier in this Atlas (unit on Well-Developed Lamination in a Black Shale-Example II). Because of the high water content of the algal mat there would be considerable volume loss upon burial during which particles could reorient into the parallel positions shown in Figure A (large arrow). It is interesting to speculate if this rock is an oil shale because hydrocarbons weren't able to migrate out owing to preferred particle orientation and resultant low permeability!

Whole Rock Composition	Wt%
Quartz	19
Plagioclase	4
Calcite	1
Siderite	4
Pyrite	3
Layer Silicates	68
Illite-Smectite	70
Illite-Mica	8
Kaolinite	18
Chlorite	4
TOC	11.3

A. SEM micrograph showing the well-developed preferred orientation which is responsible for the fissility of this shale sample. The large arrow points to a zone of parallel clay flakes. Small arrow points to a clay coated silt grain. Scale = 10 μm.

RUNDLE OIL SHALE

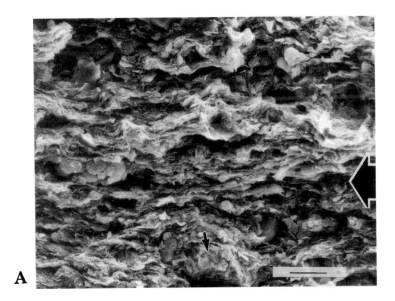

CHAPTER 9
FABRIC OF GEOPRESSURED SHALE

GEOPRESSURED SHALE ANALYSIS

This unit describes the fabric of geopressured shales in an attempt to evaluate the application of fabric analysis in understanding the origin of geopressuring. Our investigation of geopressured shales is very preliminary so we hope that the examples presented here will stimulate additional research. The term "geopressure" is used to describe pore fluid pressures which are greater than hydrostatic pressure (the pressure equal to that of a column of water extending to the surface from a given depth). Geopressures are of great importance to the oil industry because they constitute an expensive and dangerous hazard in drilling (Dickey, 1976). Geopressure usually is found within thick sequences of shales or mudstones.

A variety of processes have been proposed to account for geopressured environments. These processes are best summarized in a reprint series edited by Dutta (1987), and the reader is referred to that volume for details. According to Dutta, "mechanical compaction disequilibrium" is the principal cause of geopressuring. The process of geopressure buildup is a consequence of rapid burial of thick intervals of clay mud; thus the expulsion of pore water will be hindered during burial compaction. If the rate of loading is faster than the expulsion of pore water, then the water is trapped and ultimately supports some of the overburden load (i.e. the pore water becomes abnormally pressured).

It is also generally agreed that, at least in the Gulf of Mexico, the diagenetic conversion of smectite to illite clay through an intermediate mixed layer illite-smectite phase releases considerable bound water during burial. If this free water cannot flush through the mud pore system, then the water will begin to support the sediment load and contribute to pressure buildup.

GENERAL GEOLOGY AND COMPOSITION

Our observations have been made on shale cuttings samples from normally pressured and geopressured zones in three widely spaced wells in the Texas-Louisiana Gulf Coast area. One well is located onshore Louisiana, another is located in the High Island, Texas offshore area and the third is located in the South Padre Island, Texas offshore area.

Information on the depths, positions, and ages of normally pressured and geopressured intervals in the three wells are presented in the table below, along with depth intervals of cuttings samples and their mineralogy. Note from this table that all of the strata analyzed are Tertiary in age, and that the depth and age of the geopressured interval vary from well to well.

Presence of geopressured zones has been determined for these wells from sonic transit time logs, density logs, and mud weight data (Hottman and Johnson, 1965). The geopressured zones were recognized on well logs by deviation from a normal depth-dependent compaction trend, specifically by abnormal increase in sonic transit time (decrease in interval velocity) and decrease in rock density. Rapid increase with depth in drilling mud weight was also indicative of geopressure. Below a certain depth within the main geopressured zone in the onshore Louisiana and South Padre Island wells, sonic transit time and density values began to return with depth toward the normal compaction trend; these intervals are referred to on the table as "Below Main Geopressure".

ONSHORE LOUISIANA

Average Whole Rock Composition	Stratigraphically Above Geopressured Interval	Within Geopressured Interval	Stratigraphically Below Geopressured Interval
Quartz	56	68	40
Feldspar (total)	11	2	4
Calcite	2	5	6
Pyrite	–	–	Tr
Other	2	4	4
Layer Silicates	30	21	46
Illite-Smectite	63	53	34
Illite	2	24	41
Kaolinite	35	23	20
Chlorite	–	–	6
TOC (Wt.%)	0.6	0.4	0.4
Age	Early Miocene	Early Miocene-late Oligocene	Late Oligocene
Depth Range (ft.)	7,900-10,500	10,500-16,000	16,000-17,600
No. Samples	2	2	2

HIGH ISLAND, TEXAS OFFSHORE AREA

Average Whole Rock Composition	Stratigraphically Above Geopressured Interval	Within Geopressured Interval
Quartz	39	49
Feldspar (total)	17	14
Calcite	2	5
Pyrite	2	2
Dolomite	0	1
Layer Silicates	40	27
Illite-Smectite	53	53
Illite	32	31
Kaolinite	11	12
Chlorite	4	5
TOC (Wt.%)	0.7	0.6
Age	Pleistocene	Pleistocene-early Pliocene
Depth Range (ft.)	3,500-5,000	5,000-9,000+
No. Samples	1	3

SOUTH PADRE ISLAND, TEXAS OFFSHORE AREA

Average Whole Rock Composition	Stratigraphically Above Geopressured Interval	Within Geopressured Interval	Stratigraphically Below Geopressured Interval
Quartz	44	35	29
Feldspar (total)	7	9	11
Calcite	15	20	25
Dolomite	4	2	0
Pyrite	2	2	5
Layer Silicates	28	32	30
Illite-Smectite	53	55	43
Illite	31	27	29
Kaolinite	11	7	11
Chlorite	4	12	17
TOC (Wt.%)	0.4	0.7	0.3
Age	Pliocene-middle Miocene	Middle Miocene	Middle-early Miocene
Depth Range (ft.)	3,800-9,000	9,000-13,000	13,000-15,500
No. Samples	2	2	2

The generalized mineralogical data records diagenetic processes and reactants that have been well documented for Gulf Coast shales (Hower et al., 1976). These relationships are particularly well displayed for the onshore Louisiana well, where with increasing depth the proportion of discrete illite increases at the expense of mixed layer illite-smectite, kaolinite content decreases, and chlorite content increases. The diagenetic processes responsible for these changes include loss of bound water to adjacent pore spaces. These reactions are a function of sediment age, burial rates and temperatures.

DESCRIPTION OF SHALE FABRIC

Our observations indicate that shales above and below the main geopressured zone in these wells commonly display a finer-grained texture than geopressured shales, and a moderate to well developed clay flake orientation. By contrast, microfabric in shales from within the geopressured zones are siltier (coarser-grained) and exhibit a random particle orientation. Examples are illustrated in the SEM micrographs.

INTERPRETATION OF SHALE FABRICS

Observed variations in microfabric can be explained by the mechanical compaction disequilibrium process discussed above for the origin of geopressure. The siltier nature of geopressured shales compared to stratigraphically adjacent normally pressured shales implies higher sedimentation rates for the former, and is a factor accounting for disequilibrium.

Although silty sediments are more permeable than adjacent clayey sediments, permeabilities would still be low, fluid flow through both types of strata would be slow, and the stratigraphically adjacent less permeable clayey sediments would act to partially seal the strata from fluid expulsion. In addition, the diagenetic processes discussed above provide additional interstitial water upon burial. Thus the data presented here suggest a strong primary depositional control on geopressuring, aided by secondary diagenetic effects.

Alternately, the random, more open fabric of the geopressured shales could be a secondary effect resulting from the post-depositional pressure buildup and increase in porosity. This is unlikely, however, based on the common observation in this Atlas that silty rocks from a variety of geologic settings generally exhibit a random fabric. Clearly, future studies are required to firmly establish the relationship between origin of geopressure and fabric. We suggest here that fabric analysis should become a part of any research aimed at identifying the cause(s) of geopressuring in shale.

Geopressured Shale

A. SEM micrograph of microfabric of shale in the geopressured zone of the South Padre Island well. Fabric is characterized by numerous silt size grains (quartz?) surrounded by a matrix of randomly oriented clay flakes. The large arrow points to silt, and the smaller arrow shows a zone of random clay. Scale=10 μm.

B. SEM micrograph of microfabric of shale in the geopressured zone of the High Island well. The same fabric is apparent as was described in A. The large arrow points to silt, and the small arrow shows typical random clay microfabric. Scale=10 μm.

C. SEM micrograph of microfabric of shale in the geopressured zone of the Louisiana well. This example also shows typical random clay fabric (small arrow) and abundant silt grains (large arrow). Scale=10 μm.

Normally Pressured Shale

D. SEM micrograph of oriented microfabric of shale below the main geopressured zone in the South Padre Island Well. Scale=10 μm.

E. SEM photograph of oriented microfabric of shale above the geopressured zone in the High Island well. Scale=10 μm.

F. SEM photograph of oriented microfabric of shale just above the geopressured zone in the Louisiana Well. Scale=10 μm.

GEOPRESSURED SHALES

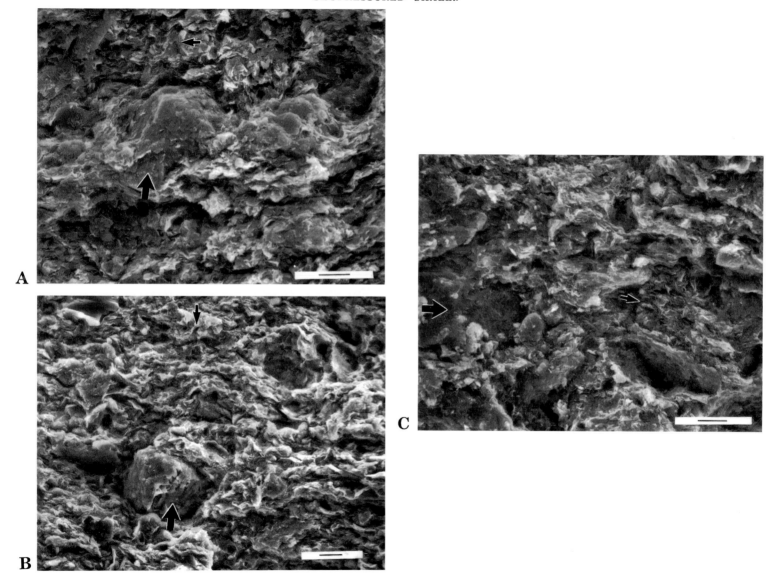

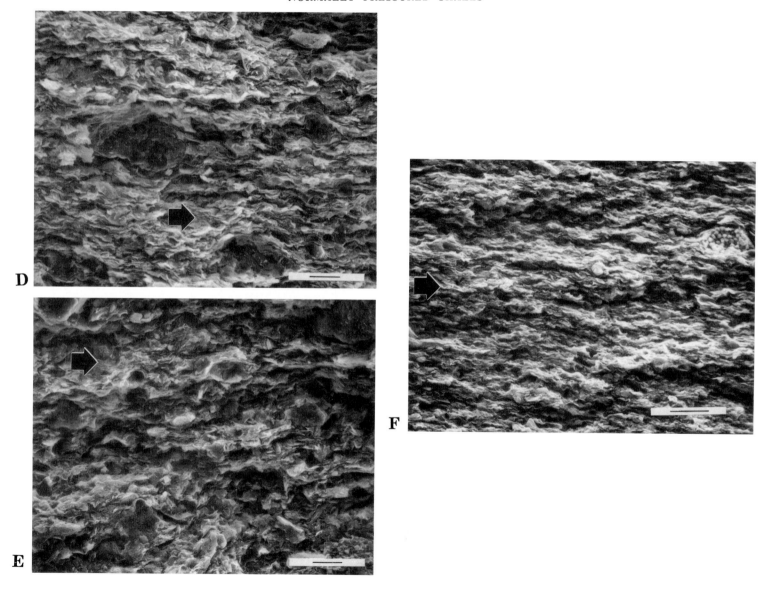

CHAPTER 10
COMPOSITION OF ARGILLACEOUS ROCKS

COMPOSITION OF ARGILLACEOUS ROCKS

Table 1 summarizes ranges and average mineralogic compositions, based on X-ray diffraction analyses, of the ninety six Ordovician-Quaternary age argillaceous rocks used in this study. Also listed are ranges and average values of Total Organic Carbon (TOC). The data in Table 1 are compiled from individual rock descriptions in the text, in addition to other analyses of rocks not given with individual rock descriptions (Table 2).

These data, though useful from the perspective of providing a mineralogic data set for argillaceous rocks of varying type, diagenesis, and age, cannot be used to quantify any systematic age or environmental-related trends. This is because the distribution of rocks by age is not uniform. In some instances, several rocks of one geologic period are from a single formation or related formations. Also, depositional environments are not uniformly represented by all rock suites.

Singer and Müller (1983) have summarized observed systematic changes in mineral composition of argillaceous rocks and diagenetic processes thought to be responsible for these changes (see their figure, this page). Specifically, with age the proportions of mixed layer illite-smectite and kaolinite clay minerals diminish and those of discrete illite and chlorite increase. K-Feldspar also diminishes with age. The data in Table 1 exhibit the same general trends. However, ranges of values for any given group of analyses often span the total range of compositions throughout the Ordovician-Quaternary time span represented by the data. Whole rock compositions are quite variable, as would be expected owing to variations in grain size, degree and type of cementation, etc. Pyrite and carbonates are considered to be diagenetic minerals, whereas quartz and feldspars are undoubtedly partially diagenetic and partially detrital in origin. K-feldspar exhibits a tendency toward becoming less abundant with geologic age, possibly reflecting dissolution upon burial. Interestingly, plagioclase, which is considered to be less chemically stable than K-feldspar, persists, and may even increase in abundance with age, suggesting it is an authigenic by-product of burial diagenesis.

The interpreted presence of authigenic quartz, feldspar, and carbonates in these argillaceous rocks indicates chemical diagenesis and cementation are common processes associated with age and burial. However, pervasive physical evidence of these processes is lacking when the rocks are examined under the SEM, which suggests that physical processes will dominate over chemical processes in formation of the ultimate fabric of the rock.

TOC values are also quite variable, and do not show any systematic change with geologic age. The amount of organic matter present in an argillaceous rock will be a function of the amount originally deposited and the nature of the depositional environment which will either allow the organic particles to become oxidized or preserved upon burial.

In summary, the compositional data reported here correspond in a general sense to compositional data reported by others. Burial diagenetic processes are invoked to explain compositional variations with geologic age and burial diagenesis. Our results indicate chemical diagenesis of argillaceous rocks does not destroy primary depositional or post-depositional compaction fabric.

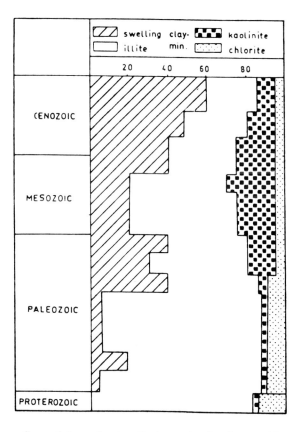

Summary figure of observed systematic changes in mineral composition as a function of age (after Singer and Müller, 1983)

TABLE 1. AVERAGE AND RANGE OF COMPOSITIONS OF ARGILLACEOUS ROCKS ACCORDING TO GEOLOGIC AGE

Age	No. Analyses	T.O.C.	Quartz	K-Feldspar	Plagioclase Feldspar	Calcite	Dolomite	Siderite
Quaternary	5	0.9 0.6- 2.1	41.0 20.0- 49.0	12.0 0.9- 17.0	--	6.4 2.0- 15.0	2.3 0.0- 8.4	--
Pliocene	4	2.8 <.1- 5.7	16.0 7.5- 24.0	6.2 1.4- 13.0	13.0 12.0- 13.0	3.5 3.1- 5.8	--	3.2 1.9- 11.0
Miocene	9	1.4 0.3- 9.9	35.0 22.0- 56.0	7.6 0.0- 11.	12.0	15.0 2.0- 25.0	1.2 0.0- 4.0	--
Oligocene	4	0.4 0.4- 0.4	54.0 40.0- 68.0	3.0 2.0- 4.0	--	5.5 5.0- 6.0	--	--
Eocene	11	3.7 0.3- 11.0	37.0 11.0- 59.0	2.1 0.0- 10.0	8.7 0.0- 24.0	4.1 0.0- 20.0	4.9 0.0- 45.0	1.8 0.0- 10.0
Cretaceous	9	2.1 0.2- 4.8	56.0 29.0- 82.0	3.8 1.0- 16.0	1.7 0.0- 4.0	3.1 0.0- 18.0	8.4 0.0- 31.0	0.1 0.0- 1.0
Jurassic	10	12.0 2.1- 46.0	24.0 14.0- 35.0	0.6- 2.0	4.8 2.0- 7.0	16.0 0.0 53.	1.8 0.0- 6.0	0.4 0.0- 4.0
Triassic	9	0.3 0.07- 0.90	47.0 33.0- 57.0	11.0 2.0- 16.0	0.7 0.0- 3.0	3.8 0.0- 23.0	4.2 0.0- 27.0	5.2 0.0- 29.0
Permian	1	0.2	28.0	4.0	8.0	--	1.0	--
Pennsylvanian	7	11.0 0.9- 29.	32.0 24.0- 40.0	0.8 0.0- 2.0	6.1 4.0- 9.0	1.4 0.0- 10.0	2.1 0.0- 15.0	3.3 0.0- 11.0
Mississippian	3	4.9 4.5- 5.8	30.0 27.0- 33.0	0.4 <1.0 1.0	3.0 3.3- 4.0	--	--	0.6 0.0- 2.0
Devonian	22	4.2 0.3- 18.6	53.0 24.0- 68.0	0.6 0.0- 4.0	0.0- 4.7 9.0	2.2 0.0- 12.0	1.5 0.0- 9.0	0.3 0.0- 2.0
Ordovician	2	1.5 1.3- 1.6	33.0 32.0- 33.0	<1.0 0.0- <1.0	6.5 6.0- 7.0	10.0 1.0- 19.0	0.5 0.0- 1.0	0.5 0.0- 1.0

TABLE 1. AVERAGE AND RANGE OF COMPOSITIONS OF ARGILLACEOUS ROCKS ACCORDING TO GEOLOGIC AGE

Age	Pyrite	Other Minerals	Total Layer Silicates	Illite-Smectite	Illite-Mica	Kaolinite	Chlorite	Others
Quaternary	5.4 2.0-19.0	--	29.0 24.0-40.0	58.0 53.0-77.0	28.0 17.0-32.0	10.0 3.4-12.0	4.3 2.4-5.0	--
Pliocene	2.0 1.0-3.0	<1.0 0.0-1.0	62.0 59.0-64.0	25.0 11.0-40.0	47.0 37.0-63.0	6.0 3.3-11.0	2.4 0.7-3.7	--
Miocene	2.0 0.0-5.0	2.5 0.0-21.0	26.0 20.0-32.0	46.0 27.0-63.0	24.0 2.0-62.0	13.0 0.0-35.0	7.7 0.0-17.0	--
Oligocene	Tr	4.0 4.0-4.0	34.0 21.0-46.0	44.0 34.0-53.0	33.0 24.0-41.0	22.0 20.0-23.0	3.0 0.0-6.0	--
Eocene	1.6 0.0-5.0	--	43.0 1.0-68.0	56 0.0-79.0	27.0 8.0-43.0	4.6 0.0-18.0	8.5 0.0-24.0	--
Cretaceous	1.6 0.0-3.0	--	29.0 11.0-63.0	21.0 9.0-39.0	28.0 18.0-49.0	40.0 30.0-54.0	11.0 0.0-25.0	--
Jurassic	12.0 10.0-21.0	--	38.0 14.0-51.0	32.0 16.0-56.0	38.0 33.0-47.0	20.0 6.0-26.0	10.0 3.0-16.0	--
Triassic	--	--	30.0 6.0-55.0	16.0 0.0-55.0	55.0 21.0-89.0	10.0 0.0-26.0	7.1 0.0-19.0	--
Permian	--	42.0	17.0	26.0	34.0	--	10.0	30.0
Pennsylvanian	3.4 0.0-8.0	--	48.0 25.0-66.0	36.0 19.0-54.0	45.0 30.0-60.0	9.6 0.0-16.0	9.4 5.0-14.0	--
Mississippian	5.3 5.0-6.0	--	59.0 57.0-62.0	27.0 21.0-30.0	58.0 57.0-60.0	6.0 4.0-10.0	9.0 9.0-9.0	--
Devonian	3.7 0.0-7.0		47.0 12.0-64.0	7.7 0.0-49.0	76.0 42.0-93.0	2.3 0.0-9.0	14.0 2.0-25.0	--
Ordovician	3.5 3.0-4.0		46.0 35.0-56.0	11. 0.0-21.0	66.0 58.0-74.0	--	23.0 21.0-25.0	--

125

TABLE 2. MISCELLANEOUS COMPOSITIONAL ANALYSES OF ROCKS DESCRIBED ELSEWHERE IN THE ATLAS

	Quartz	K-Feldspar	Plagioclase	Calcite	Dolomite	Siderite	Pyrite	Total Layer Silicate	Illite-Smectite	Illite-Mica	Chlorite-Smectite	Chlorite	TOC	Age
Mecca Quarry Shale Member	36	--	6	--	--	2	4	52	54	30	--	7	6.6	Penn.
Mecca Quarry Shale Member	32	--	5	--	--	1	8	54	24	60	--	4	18.4	Penn.
Canton Shale Member	25	<1	6	--	--	11	1	57	31	44	--	12	1.1	Penn.
Cashaqua Shale Member	33	1	5	7	2	--	2	50	--	83	--	17	0.7	Dev.
Cashaqua Shale Member	37	<1	4	5	--	--	1	52	--	72	--	23	0.3	Dev.
Java Formation	39	<1	4	2	1	2	2	50	--	85	--	14	NA	Dev.
Genesee Shale Member	33	1	4	12	4	--	4	44	--	82	--	16	2.5	Dev.
Jet Rock Formation	33	<1	4	4	--	--	11	48	26	35	--	12	2.2	Jur.
Energy Shale	24	<1	8	--	--	1	<1	66	42	38	--	12	0.9	Penn.
Rhinestreet Shale Member	50	1	9	--	--	--	5	35	--	71	--	25	4.3	Dev.
Utica Shale	33	--	7	1	--	--	3	56	21	58	--	21	1.3	Ord.
Utica Shale	32	<1	6	19	--	1	4	35	--	74	--	25	1.6	Ord.
Chagrin Shale Member	28	<1	3	--	--	--	4	64	--	85	--	7	0.5	Dev.
Middlesex Shale Member	42	1	6	--	--	--	3	49	--	77	--	21	2.6	Dev.
Huron Shale Member	36	<1	5	--	--	--	5	54	14	74	--	10	1.0	Dev.
Huron Shale Member	58	--	5	--	--	--	6	31	--	93	--	5	6.5	Dev.

TABLE 2. MISCELLANEOUS COMPOSITIONAL ANALYSES OF ROCKS DESCRIBED ELSEWHERE IN THE ATLAS

	Quartz	K-Feldspar	Plagioclase	Calcite	Dolomite	Siderite	Pyrite	Total Layer Silicate	Illite-Smectite	Illite-Mica	Kaolinite	Chlorite	TOC	Age
Cleveland Shale Member	55	1	6	--	--	--	3	35	22	70	5	2	8.6	Dev.
Roof Shale over Springfield V Coal	33	1	5	--	--	--	7	55	43	46	7	5	29.0	Penn.
Pen Yan Shale Member	47	--	6	--	--	--	3	45	--	77	4	19	4.8	Dev.
Bedford Formation	33	1	4	--	--	--	6	57	21	60	10	9	5.8	Miss.
Sunbury Formation	29	<1	3	--	--	2	5	62	30	57	4	9	4.5	Miss.
Canton Shale Member	32	2	9	--	--	8	--	49	19	54	16	12	1.1	Penn.
New Albany Formation	24	1	4	--	4	--	7	60	27	53	9	11	18.6	Dev.
Wilcox Group	53	--	12	19	--	--	1	53	50	37	3	10	0.3	Eoc.
Wilcox Group	42	--	3	--	1	2	1	51	79	11	--	10	1.3	Eoc.
Pico Formation	17	6	13	--	--	1	--	62	29	63	5	2	--	Plio.

TABLE 2. MISCELLANEOUS COMPOSITIONAL ANALYSES OF ROCKS DESCRIBED ELSEWHERE IN THE ATLAS

	Quartz	K-Feldspar	Plagioclase	Calcite	Dolomite	Siderite	Pyrite	Total Layer Silicate	Illite-Smectite	Illite-Mica	Kaolinite	Chlorite	TOC	Age
Bituminous Shale Formation	21	--	2	14	1	--	11	51	31	37	25	7	3.1	Jur.
Jet Rock Formation	17	2	6	29	5	--	14	28	38	35	18	9	7.8	Jur.
Gray Shale Formation	35	1	7	--	--	--	10	47	16	43	25	16	2.1	Jur.

CHAPTER 11
CONCLUSIONS

SUMMARY

Our analysis of the macro- and micro-fabrics of numerous argillaceous rocks demonstrates one clear conclusion - shale and mudstone fabrics are variable and that the variations may be attributed both to geologic processes operating on the sediment before, during, and after burial and to the environment in which the sediment is deposited. Many scientists often report their findings with the disclaimer that "more work needs to be done." We are no exception, because shale and mudstone research clearly is in its infancy. This atlas is an attempt to describe and discuss a group of rocks long ignored by geologists. Our conclusions are presented below.

1. The investigation of fabrics should be added to the list of methods used in a complete study of argillaceous rocks, occupying a place along with mineralogical, geochemical and other methods.

2. A complete fabric analysis of argillaceous rocks should include x-radiography, petrography, and scanning electron microscopy. The significance of the results from any one of these techniques often becomes more obvious when supported by data from another technique.

3. Careful sample preparation is very important. For example, at the initial stages of a study, a random chip from a large sample for SEM analysis may not provide any significant specific information. One should first evaluate fabric features from x-radiographic and petrographic analysis and then use that data to properly choose a sample for SEM examination. The world of the millimeter should be investigated before the world of the micrometer.

4. It is possible to classify shales and mudstones by their macrofabric as seen in x-radiographs. Lamination is either well developed or indistinct in shales and is a significant characteristic of this rock type. Lamination is lacking or greatly distorted in mudstones.

5. Petrographic macrofabric signatures vary among shales and mudstones. For example, we show four categories of black shales. Some are finely laminated and others are thickly laminated, whereas others show wavy or lenticular lamination. The different formative processes require future research.

6. Black shale fabrics also vary at the micrometer level. Microfabric variations in all types of shale are influenced by the relative abundance of clay, particulate organic matter, and silt grains. In describing a black shale fabric from an SEM micrograph, one may call it: a) organic hash, b) organic-clayey, c) clayey, or d) silty.

7. Our SEM analysis supports the observations of others in showing preferred or moderately preferred platy particle orientation in shale and randomness in mudstone.

8. The exact mechanism responsible for particle orientation is a subject for future research by sedimentologists, particularly those with a background in colloid and surface science. However, we offer some working hypotheses here which should aid in environmental interpretations. We have found that a random fabric in mudstones is attributed to bioturbation (note: a bioturbation interpretation should also be supported by x-radiographic and petrographic evidence) or preservation of a primary flocculated clay fabric. Preferred orientation in shales results most often from compaction and reorientation of flocculated clay, but may also form from shearing due to bottom currents or by deposition of dispersed clay. Our results indicate it is not always easy to interpret which mechanism is dominant in forming a fabric; however, in many cases a good first approximation of cause is possible when combined with other geologic evidence.

9. Shales and mudstones from widely varying sedimentary environments exhibit different fabrics because the dominant sedimentary processes are different in certain environments. Some processes (such as bioturbation, bottom flowing currents, slow sediment accumulation, etc.) may dominate in one environment or under certain sedimentary conditions and not in another, resulting in a unique environmental signature upon sediment fabric. On the other hand, some sedimentary processes are common in numerous environments, so that the fabric can only be used to interpret the process and not the specific environment. An example of the latter is the randomness of flocculated clay in basinal turbidite and evaporite sequences, which is indicative of deposition and preservation of primary clay floccules.

10. We have only just begun to investigate the importance of fabric analysis in evaluating hydrocarbon source rocks and oil shales. Initial results from marine argillaceous source rocks suggest an association of an open network of micropores (micrometers in scale) with an "organic hash" fabric in organic-rich black shales. It is possible that this open network could facilitate primary migration of hydrocarbons if the fabric is preserved upon burial down into the oil generative window. Our data are less conclusive, however, about the role of fabric in lacustrine oil shales simply because the fabrics of these rock types vary considerably.

11. Although it is often assumed that chemical diagenesis and cementation occur upon burial of argillaceous rocks, we have found relatively little

observational evidence that these processes are sufficiently pervasive to overprint primary fabrics. True cements appear to be volumetrically insignificant in these rocks.

12. Based upon limited analysis, we have observed that geopressured shales in the Gulf of Mexico are relatively coarse grained (silty) and exhibit random microfabric. By contrast, stratigraphically adjacent shales are finer-grained and exhibit oriented microfabric. These differences are consistent with the interpretation that geopressure develops primarily under conditions of relatively high sedimentation rate and burial compaction of mud.

These twelve statements above indicate our current knowledge of argillaceous rock fabrics. As we prepared this Atlas it became apparent that there are areas which require future investigation. Some of these are listed below in the form of questions or as suggestions for future study.

1. There is a need for a quantitative scheme to classify fabrics observed in the SEM. Such a scheme should provide a rapid way to measure particle orientation from a scanning electron micrograph.

2. At what stage in the lithification process does shale fabric form? At what depth? Is shale fabric formation related to rate of sediment accumulation? Study of cores of recent sediment would provide valuable insights into these questions. Duplication of depositional conditions in the laboratory and analysis of resulting fabrics would also prove useful.

3. There still is a need for determining the fabric of flocculated sediment in suspension before it reaches the seafloor or lake bottom. *In situ* collection and preparation techniques need to be developed in order to study undisturbed samples like "marine snow" or clay aggregates during sedimentation, at the sediment-water interface and at shallow depths of burial.

4. What actually are the mechanisms responsible for reorientation of flocculated bottom sediment due to shearing of bottom flowing currents? How does a fabric formed by this mechanism differ from that produced by compaction?

5. What are the gross fabric characteristics of a mass of authigenic clay compared to undisturbed primary flocculated or bioturbated fabric?

6. There is a need to study further the various fabrics of hydrocarbon source rocks and oil shales. Here an integrated approach, which couples geochemical analysis and fabric analysis, would be most valuable in providing clues to some aspects of hydrocarbon generation and migration.

7. What processes are responsible for forming the various types of black shale laminae seen in thin-section? For example, what causes wavy or lenticular lamination?

8. What is the significance of the shape and distribution of micropores in hydrocarbon source rocks? Is one type of fabric (e.g. organic-hash) more common in source rocks? Does the open fabric of organic-rich rocks promote primary hydrocarbon and other fluid migration?

9. More fabric analysis is needed to determine why an oil shale retains a kerogen content sufficiently high enough to produce hydrocarbons upon laboratory processing. Why haven't the hydrocarbons migrated from this type of rock like other hydrocarbon source rocks? Is the answer locked away in fabric analysis?

10. What chemical diagenetic processes are associated with argillaceous rocks and how pervasive are these?

11. More research is needed to test our limited observations that geopressured shales have distinctly different microfabrics from stratigraphically adjacent shales. Also, the application of microfabric analysis to understanding the origin of geopressuring needs to be evaluated.

REFERENCES

Amstutz G C, Park W C, Schot E H and Love L G (1967) Orientation of framboidal pyrite in shale. Mineralium Deposits, 1:317-321.

Atkinson C D, Trumbly P N and Kremer M C (1988) Sedimentology and depositional environments of the Ivishak Sandstone, Prudhoe Bay Field, North Slope, Alaska. In: Lomando A J and Harris P M (eds.) Giant Oil and Gas Fields, A Core Workshop. Society of Paleontologists Mineralogists Core Workshop 12:561-614.

Bates R L and Jackson J A (1987) Glossary of Geology, American Geological Institute, Falls Church, Va., 788 pp.

Bennett R H, Bryant W R and Keller G H (1979) Clay fabric and related pore geometry of selected submarine sediment, Scanning Electron Microscopy 1979. I. pp 519-527 and 424.

Bennett R H and Hulburt M H (1986) Clay Microstructure. IHRDC Press, Boston, 160 pp

Berner R A (1969) The synthesis of framboidal pyrite. Economic Geology 64:383-384.

Berner R A (1970) Sedimentary pyrite formation. American Journal Science 268:1-23.

Bitterli P (1963) Aspects of the genesis of bituminous rock sequences. Geologic en Mijnbouw 42:183-201.

Blatt H, Middleton G and Murray R (1980) Origin of Sedimentary Rocks, 2nd Edition. Prentice-Hall, Englewood Cliffs, New Jersey 782 pp.

Boles J R and Franks S G (1979) Clay diagenesis in Wilcox sandstones of southwest Texas: Implications of smectite diagenesis on sandstone cementation. Journal Sedimentary Petrology 49:55-70.

Bouma A N (1969) Methods for the study of sedimentary structures. Wiley-Interscience, New York 458 pp.

Bradley J S (1975) Abnormal formation pressure. American Association Petroleum Geologists Bulletin 59:957-973.

Bromley R G and Ekdale A A (1984) Chrondrites: A trace fossil indicator of anoxia in sediments. Science 224:872-874.

Bromley R G and Ekdale A A (1986) Composite ichnofabrics and tiering of burrows. Geological Magazine 123:59-65.

Burst J F (1967) Diagenesis of Gulf Coast clayey sediments and its possible relationship to petroleum migration. American Association Petroleum Geologists Bulletin 53:73-93.

Byers C W (1977) Biofacies patterns in euxinic basins: a general model. In: Cook H E and Enos P (eds.) Deep-water Carbonate Environments. Society Economic Paleontologists Mineralogists, Special Publication 25:5-17.

Calvert S W and Veevers J J (1962) Minor structures of unconsolidated marine sediments revealed by x-radiography. Sedimentology 1:287-295.

Clifton H E (1966) X-ray radiography with x-ray diffraction equipment. Journal Sedimentary Petrology 36:620-635.

Cluff R M (1980) Paleoenvironment of the New Albany Shale Group (Devonian-Mississippian) of Illinois. Journal Sededimentary Petrology 50:767-780.

Cole R D and Pickard M D (1975) Primary and secondary sedimentary structures in oil shale and other fine-grained rocks, Green River Formation (Eocene), Utah and Colorado, Utah Geology 2:49-67.

Comer J B and Hinch H H (1987) Recognizing and quantifying expulsion of oil from the Woodford Formation and age-equivalent rocks in Oklahoma and Arkansas. American Association Petroleum Geology Bulletin 71:844-858.

Curtis C D (1980) Diagenetic alteration in black shales. Journal Geological Society London 137:189-194.

Dean W E, Arthur M A and Stow D A V (1984) Origin and geochemistry of Cretaceous deep-sea black shales and multicolored claystones, with emphasis on deep-sea drilling project site 530, southern Angola Basin. In: Hay W W and Sibuet J C (eds.) Initial reports of the Deep Sea Drilling Project, V. LXXV, U.S. Government Printing Office, Washington, D.C. pp 819-844.

Dickey P A (1976) Abnormal formation pressure: Discussion. American Association Petroleum Geologists Bulletin 60:1124-1127.

Drake D W (1971) Suspended sediment and thermal stratification in Santa Barbara Channel, California Deep Sea Research 18:763-769.

Dunbar C O and Rodgers J (1957) Principles of Stratigraphy, Wiley and Son, Inc. New York, 356 pp.

Dutta N C (1987) Geopressure. Society Exploration Geophysics, Geophysics Report Series 7, 365 pp.

Ece O I (1987) Petrology of the Desmoinesian Excello black shale of the Midcontinent region of the United States. Clays and Clay Minerals 35:262-270.

Ekdale A A (1985) Paleoecology of the marine endobenthos. Paleogeography, Paleoclimatology, Paleoecology 50:63-81.

Ettensohn F R (1985) Controls on development of Catskill Delta complex basin facies, in the Catskill Delta. In: Woodrow DL and Sevon WD (eds) Geological Society America Special Paper 201:65-78.

Folk R L (1962) Petrography and origin of the Silurian Rochester and McKenzie shales, Morgan County, West Virginia. Journal Sedimentary Petrology 32:539-578.

Folk R L (1968) Petrology of sedimentary rocks Hemphill's Bookstore, Austin, Texas, 170 pp.

Gad M A, Cott J A and LeRiche H H (1969) Geochemistry of the Whitbian (Upper Lias) sediments of the Yorkshore coast. Proceedings Yorkshire Geological Society 37:105-139.

Gallois R W (1972) Coccolith blooms in the Kimmeridge clay and origin of North Sea Oil. Nature 259:473-475. London.

Gaynor G C and Scheihing M H (1988) Shelf depositional envirinments and reservoir characteristics of the Kuparuk River Formation (Lower Cretaceous), Kuparuk Field, North Slope, Alaska. In: Lomando A J and Harris P M (eds.) Giant Oil and Gas Field, a Core Workshop. Society Economic Paleontologists Mineralogists Core Workshop 1:333-390.

Grim R E (1962) Applied Clay Mineralogy. McGraw-Hill, New York, 422 pp.

Hallam A (1967) The depth significance of black shales with bituminous laminae, Marine Geology 5:481-493.

Hallam A (1975) Jurassic environments. Cambridge University Press, Cambridge, 269 pp.

Hamblin W K (1962) X-ray radiography in the study of structures in homogeneous sediments. Journal Sedimentary Petrology 32:201-210.

Handford C R (1981) Coastal sabkha and salt pan depositon of the Lower Clear Fork Formation (Permian), Texas. Journal Sedimentary Petrology 51:761-778.

Harwood R J (1977) Oil and gas generation by laboratory pyrolysis of kerogen. American Association Petroleum Geologists Bulletin 61:2082-2102.

Hattin D E (1975) Petrology and original of fecal fellets in Upper Cretaceous strata of Kansas and Saskatchewan. Journal Sedimentary Petrology 45:686-696.

Heckel P H (1977) Origin of phosphatic black shale facies in Pennsylvanian cyclothems of Mid-Continent North America. American Association Petroleum Geologists Bulletin 61:1045-1068.

Hinch H H (1980) The nature of shales and the dynamics of hydrocarbon expulsion in the Gulf Coast Tertiary section. In: Roberts W H, III and Cordill R J (eds.) Problems of petroleum migration, AAPG Studies in Geology No. 10 1-18 pp.

Hosterman J W and Whitlow S I (1983) Clay mineralogy of Devonian shales in the Appalachian Basin. United States Geological Survey Professional Paper 1298, 31 pp

Hottman C E and Johnson R K (1965) Estimation of formation pressure from lug-derived shale properties. Journal Petroleum Technology June, pp 717-722

Howarth M K (1962) The Jet Rock Series and the Alum shale Series of the Yorkshire coast. Proceedings Yorkshire Geological Society 33:381-422.

Hower J, Eslinger E V, Hower M E and Perry E A (1976) Mechanism of burial metamorphism of argillaceous sediment: 1, Mineralogical and chemical evidence. Geological Society America Bulletin 87:725-737.

Hunt S M (1979) Petroleum geochemistry and geology. Freeman and Co., San Francisco, 617 pp.

Hutton A C, Kantsler A J, Cook A C and McKirdy D M (1980) Organic matter in oil shales. Australian Petroleum Exploration Association Journal 20:44-67.

Ingram R L (1953) Fissility of mudrocks. Geological Society America Bulletin 65:869-878.

Isaacs C M (1981) Lithostratigraphy of the Monterey Formation, Goleta to Point Conception, Santa Barbara coast, California. In: Isaacs C M (ed.) Guide to the Monterey Formation in the California coastal area, Ventura to San Juis Obispo. Pacific Section American Association Petroleum Geologists field guide pp 9-24.

Javor B J and Mountjoy E W (1976) Late Proterozoic microbiota of the Mietta Group, southern British Columbia. Geology 4:111-119.

Jordan D W (1985) Trace fossils and depositional environments of Upper Devonian black shales, east-central Kentucky U.S.A. In: Curran H A (ed) Biogenic structures: their use in interpreting depositional environments, Society Economic Paleontologists Mineralogists Special Publication 35:279-298.

Kalliokoski J (1974) Pyrite Framboid: animal, vegetable, or mineral? Geology 2:26-27.

Lewan M D (1979) Laboratory classification of very fine-grained sedimentary rocks, Geology 6:745-748.

Lindner A W and Dixon D A (1976) Some aspects of the geology of the Rundle oil shale deposit, Queensland, Australia. Petroleum Exploration Association Journal 12:165-172.

Love L G (1967) Early diagenetic iron sulphide in recent sediments of the Wash England. Sedimentology 9:327-352.

Lundegard P D and Samuels N D (1980) Field classification of fine-grained sedimentary rocks. Journal Sedimentary Petrology 50:781-786.

Morris K A (1979) A classification of Jurassic marine shale sequences: an example from the Toarcian (Lower Jurassic) of Great Britain. Palaeogeography, Palaeoclimatology, Palaeoecology 26:117-126.

Morris K A (1980) Comparison of major sequences of organic-rich deposition in the British Jurassic. Journal Geological Society London 137:157-170.

Nuhfer E B, Vinopal R J and Klanderman D S (1979) X-radiograph atlas of lithotypes and other structures in the Devonian shale sequence of West Virginia and Virginia: METC/CR-79/27, NTIS, Springfield, Virginia 45 pp.

Nuhfer E B (1981) Mudrock fabrics and their significance: discussion. Journal Sedimentary Petrology 51:1027-1029.

O'Brien N R and Slatt R M (1988) Determining sedimentary processes and environments of argillaceous rocks by fabric analysis techniques. American Association Petroleum Geologists Bulletin 72:231.

O'Brien N R (1968) Electron microscope study of black shale fabric, Naturwissenschaften 10:490-491.

O'Brien N R (1981) SEM study of shale fabric-a review. Scanning Electron Microscopy I 569-575 pp.

O'Brien N R (1987) The effects of bioturbation on the fabric of shale. Journal Sedimentary Petrology 57:449-455.

O'Brien N R (1989) Origin of lamination in middle and upper Devonian black shales, New York. Northeastern Geology 11:159-165.

O'Brien N R, (1990) Significance of lamination in the Toarcian (Lower Jurassic) shales from Yorkshire, Great Britain. Sedimentary Geology 67, Elsevier Science Publishers, the Netherlands.

O'Brien N R, Nakazawa K, and Tokuhashi S (1980) Use of clay fabric to distinguish turbiditic and hemipelagic siltstones and silts. Sedimentology 27:47-61.

Picard M D (1971) Classification of fine-grained sedimentary rocks. Journal Sedimentary Petrology 41:179-195.

Pierce J W (1976) Suspended sediment transport at the shelf break and over the outer margin, In: Stanley D J and Swift D J P (eds) Marine sediment transport and environmental management. Wiley and Sons, Inc., N.Y. pp 437-458.

Piper D J W (1972) Turbidite origin of some laminated mudstones. Geological Magazine 109:115-126.

Piper D J W (1978) Turbidite muds and silts on deep-sea fans and abyssal plains, In: Stanley D J and Kelling G (eds) Sedimentation in submarine canyons, fans, and trenches Dowden, Hutchinson, and Ross, Stroudsburg, PA. pp 163-176.

Pisciotto K A and Garrison R E (1981) Lithofacies and depositional environments of the Monterey Formation, California, In: Garrison R E, Douglas R G, Pisciotto K E, and Ingle J C (eds) The Monterey Formation and related siliceous rocks of California. Pacific Section Society Economic Paleontologists Mineralogists Special Publication pp 97-122.

Porter K G (1984) Alternate fate of planktonic detritus: organic deposition and the geological record. Bulletin Marine Science 35:587-600.

Potter P E, Maynard J B and Pryor W A (1980) Sedimentology of Shale. Springer-Verlag, New York 310 pp.

Powers M C (1967) Fluid-release mechanisms in compacting marine mudrocks and their importance in oil exploration. American Association Petroleum Geologists Bulletin 51:1240-1254.

Presley M W and McGillis K A (1982) Coastal evaporite and tidal-flat sediments of the Upper Clear Fork and Glorieta Formations, Texas Panhandle. Report Investigation 115, Bureau Economic Geology University of Texas at Austin 50 pp.

Pryor W A (1975) Biogenic sedimentation and alteration of argillaceous sediments in shallow marine environments. Geological Society America Bulletin 86:1244-1254.

Reif D W and Slatt R M (1979) Red bed members of the lower Triassic Moenkopi Formation, southern Nevada: Sedimentology and paleogeography of a muddy tidal flat deposit. Journal Sedimentary Petrology 49:869-890.

Rickard D T (1970) The origin of framboids. Lithos 3:269-293.

Riegel W, Loh H, Maul B and Prauss M (1986) Effects and causes in a black shale event - The Toarcian Posidonia shale of NW Germany, In: Walliser O (ed) Lecture Notes in Earth Sciences, Global Bio-Events. Springer-Verlag, Berlin 8:267-276.

Rieneck H E and Singh I B (1978) Depositional Sedimentary environments, with reference to terrigenous clastics. Springer-Verlag, Berlin 549 pp.

Robinson W E (1976) Origin and characteristics of Green River oil shale, In: Yen T F and Chilingarian G V (eds) Amsterdam Oil Shale, Elsevier Scientific Company. Developments in Petroleum Science 5:61-79.

Rosen M R (in press) Mineralogic and stable isotopic constraints on the hydrologic evolution of Bristol Dry Lake, California, U.S.A. Palaeogeography, Palaeoclimatology, Palaeoecology.

Schutter S R and Heckel P H (1985) Missourian (Early Late Pennsylvanian) climate in Midcontinent North America. International Journal Coal Geology 5:111-140.

Schwietering J F (1979) Devonian shales of Ohio and their eastern and southern equivalents METC/CR-79/2, NTIS, Springfield, VA 68 pp.

Shanks W C, Seyried W E, Meyer W C and O'Neil J J (1976) In: Yen T F and Chilingar G V (eds) Developments in Petroleum Science. New York, Elsevier, 5:81-102.

Singer A and Müller G (1983) Diagenesis in argillaceous sediments In: Larsen G and Chilingar G V (eds) Diagenesis in sediments and sedimentary rocks, 2, Developments in Sedimentology 25B, Elsevier, pp 115-212.

Slatt R M, Boak J M, Goodrich G T, Lagoe M B, Vavra C L, Bishop J M and Zucker S M (1988) In: Lomando A J and Harris P M (eds) Depositional facies, paleoenvironments, reservoir quality, and well log characteristics of Mio-Pliocene deep water sands, Long Beach Unit, Wilmington Field, California, Giant Oil and Gas Fields. Society Economic Paleontologists Mineralogists Core Workshop 12, 7:31-88.

Slatt R M and Thompson P R (1985) Submarine slope mudstone facies, Cozy Dell Formation (middle Eocene), California. Geo-Marine Letters 5:39-45.

Smith N D and Syvitski J P M (1982) Sedimentation in a glacier-fed lake: the role of pelletization on deposition of fine-grained sediments. Journal Sedimentary Petrology 52:503-513.

Spears D A (1976) The fissility of some Carboniferous shales. Sedimentology 23:721-725.

Spenser R J, Baedecker M J, Eugster H P, Forester R M, Goldhaber M B, Jones B F, Kelts K, McLenzie J, Madsen D B, Rettig S L, Rubin M and Bowser C J (1984) Great Salt Lake, and precursors, Utah: the last 30,000 years. Contributions Mineralogy Petrology 86:321-334.

Stanley D J (1983) Parallel laminated deep-sea muds and coupled gravity -- hemipelagic settling in the Mediterranean. Smithsonian Contributions Marine Science No. 19, 19 pp.

Stow D A V and Bower A J (1980) A physical model for the transport and sorting of fine-grained sediment by turbidity currents. Sedimentology 27:31-46.

Stow D A V and Shanmugan G (1980) Sequence of structures in fine-grained turbidites: Comparison of recent deep-sea fan and ancient flysch sediments. Sedimentary Geology 25:23-42.

Sullivan K L (1985) Organic facies variation of the Woodford Shale in western Oklahoma. Shale Shaker, Feb./March 1985 pp 76-89.

Thiessen R (1925) Microscopic examination of Kentucky oil shales, In: Thiessen R, White D and Crouse C S (eds) Oil Shales of Kentucky. Kentucky Geological Survey Series VI, 21:1-48.

Thompson S L (1985) Ferron sandstone member of the Mancos Shale, Turonian mixed energy deltaic system, Austin. University of Texas, Master's thesis 165 pp.

Thompson S L, Ossian C R and Scott A J (1986) Lithofacies, inferred processes and log response characteristics of shelf and shoreface sandstones, Ferron sandstone, central Utah, Modern and Ancient Shelf Clastics, In: Moslow T F and Rhodes E G (eds) A Core Workshop. SEPM core Workshop No. 9, pp 325-361.

Thompson R W (1968) Tidal flat sedimentation on the Colorado River Delta, northwestern Gulf of California. Geological Society America Memoir 107, 133 pp.

Tissot B P, Durand B, Espitalie J and Combaz A (1974) Influence of the nature and diagenesis of organic matter in the formation of petroleum. American Association Petroleum Geologists Bulletin 58:499-506.

Traverse A (1988) Paleopalynology. Allen and Unwin, Winchester, Massachusetts 600 pp.

Tyson R V, Wilson R C L and Downie C (1979) A stratified water column environmental model for the type Kimmeridge Clay. Nature 277:377-380.

van Olphen H (1963) An introduction to clay colloid chemistry, Interscience Publishers. Wiley and Sons, New York and London 301 pp.

Welte D (1974) Recent advances in organic geochemistry of humic substances and kerogen, In: Tissot B and Bienner F (eds), Advances in Organic Geochemistry. Editions Technip, Paris, pp 3-14.

Welton J E (1984) SEM Petrology Atlas. American Association of Petroleum Geologists Methods in Exploration Series 237 pp.

Woodrow D L (1985) Paleogeography, paleoclimate and sedimentary processes of the Devonian Catskill Delta in the Catskill Delta, In: Woodrow D L and Sevon W D The Catskill Delta. Geological Society American Special Paper 201:51-64.

Wunderlich F (1967) Reinblattrige Wechselschichtung und Gezeitenschichtung: Senckenberg, Lethaea 48:337-343.

Yen T F and Chilingar G V (1976) Developments in Petroleum Science. New York, Elsevier 5:292 pp.

INDEX

abnormally pressured shale, *see* geopressured shale
aerobic environment, 46, 58, 60, 66, 72, 76, 78
Alaska, 62, 76
algal mat, *see* microbial mat
alginite, 112
anaerobic environment
   cause of, 90
   example of, 40-54, 58, 60, 88, 90, 100, 104, 106, 111-112
analysis techniques, 8-10
anoxic, *see* anaerobic
argillaceous, definition, 2
argillaceous rock mineralogy, 123-128
Atlantic Ocean, 34
authigenic cement, 93
Australia, 112

basic environment, 60, 100, 104, 106
bioturbation
   model in shale, 55
   shown in SEM, 25, 27, 49, 51, 53, 59, 60, 75, 77-80
   shown in thin-section, 22, 51, 60, 75, 77, 80
   shown in x-radiographs, 13, 49, 51, 53, 55, 59
bituminous shale, *see* black shale
black shale
   fabric variations in SEM, 28
   lamination, significance of, 40-44, 88
   lamination type, 12, 16, 18-21
   petrographic classification, 16-17
   pyrite framboids, 33
bottom erosion, evidence of, 47, 67, 71-73, 85
bottom flowing currents, 46, 61, 67, 71, 76-77, 84-85, 93

California, 64, 82, 86, 106
classification systems, 5-6
   petrography, 16-17
   x-radiography, 12-13
clay, definition, 2
coccoliths, 100-102
Colorado, 110

compaction, 91, 117
composition of argillaceous rocks, 123-128
convolute bedding, 66-68, 72-73
cross-bedding, 23, 66-67
Cretaceous, 12, 76-78
current bedding, 23, 67, 71-73, 76-77, 79, 84-85
cut-and-fill, 47, 66-67, 85

definitions, 2-3
deltaic environment, 78
detached turbid layer, 88-90
Devonian, 12, 16, 18-23, 26-28, 32, 36, 48, 52, 60, 84, 88, 93, 104
diagenesis, 116-117
diatoms, 108
dispersed clay, model, 24
dolomite, 110-111
dysaerobic environments, 46-50, 60

energy dispersive x-ray analysis (EDX), 108, 111
England, 12, 16, 19-20, 22-24, 32, 34, 40-44, 50, 93, 100
environment, sedimentary, *see the specific environment*
episodic sedimentation, 84, 88
evaporites, 64-66, 110
   clay fabric in SEM, 27, 65
   environment, 64
   gypsum, 69

fabric, definition, 2
fecal pellets, 34-35
fissility
   definition, 2-3
   in SEM, 25, 44, 60
flocculation
   examples, 46, 64, 67-69, 86, 92, 104, 106, 109
   model of forming, 24, 92
floodplain environment, 62, 78
fluvial environment, 62

foraminifera, 108
framboids, *see* pyrite framboids

gel strength, 93
geopressured shale, 116-118
graded bedding, 23, 84-85
gray shale, 22-23, 48-51, 58
gypsum, 66, 69

hydrocarbon source rock, 76, 86, 98-100, 104-106, 110-112

Illinois, 12, 16, 21-22, 32
Indiana, 21, 28, 34
Iowa, 21, 32, 58

Jurassic, 12, 16, 19-24, 32-34, 40-44, 50, 93, 100

kerogen, 98, 110

lacustrine, 64, 110-112
   brackish, 112
   freshwater, 112
   saline, 110
lamination
   bottom currents, cause of, 71, 76-78, 81
   definition, 2-3
   detached turbid layer, cause of, 88
   environmental significance, 15, 40, 42, 46
   organic variation, in, 44
   petrographic classification, 16
   turbidity currents, 46
   types, 12, 16
   x-radiography classification, 12-13
Louisiana, 116

marine
  environment, 40-52, 58-60, 66, 72, 76, 82-88, 100, 104-106
  snow, 106
megacyclothem, 58
microbial mat, 42, 112
mineralogy of argillaceous rocks, 123-128
  by clay composition, 123-128
  by geologic age, 123-128
  by non-clay composition, 123-128
  examples in atlas, 123-128
Mississippian, 24, 46, 104
mudstone
  definition, 2-3
  fabric in SEM, 27, 62, 65, 83, 87, 106
  origin of particle orientation, 26, 82, 86, 106

non-fissile argillaceous rock, 3
non-laminated argillaceous rock, 3
Nevada, 66
New York, 12, 16, 18-19, 22-23, 26, 28, 32, 60, 88, 93

Ohio, 16, 19-20, 23-24, 36, 46, 48, 84, 93
oil shale, *see* hydrocarbon source rock
Oklahoma, 104
Ordovician, 18, 28
organic hash, 28, 58, 103, 105, 112
organic shale, *see* black shale

paleosol, 62
palynomorphs, 36-37, 103, 105
particle orientation, origin of,
  preferred, 24
  random, 24
parting types, 4
Pennsylvania, 12
Pennsylvanian, 12, 16, 21-22, 28, 32, 34, 58
permeability, 117

Permian, 26
petrography
  bioturbation, 22, 51, 59, 61, 74, 80
  black shale classification, 16-17, 18-21
  cross-bedding, 23
  current bedding, 23, 47, 85
  graded bedding, 23
  miscellaneous shale features, 23
  pyrite, 23
  techniques, 8-9
phosphatic facies, 106
Pleistocene, 26, 64, 93
pollen, 36-37
porosity, 117
preferred clay orientation, 25, 33, 41-49, 53, 61, 68, 70, 74, 77, 81, 85, 89, 94, 113, 118, 120
process
  aggregation, 106
  bioturbation, 12, 22, 26, 34, 48, 50, 58, 61, 73, 76, 78
  compaction, 36, 92
  convolute bedding, 66, 72
  cross-bedding, 23
  current lamination, 23, 66, 72, 76, 78
  cut-and-fill, 84
  detached turbid layer, 88
  dispersion, 24, 112
  fecal pellets, 34
  fine lamination, 12, 23, 61, 66, 88
  flocculation, 24, 26, 64, 86, 92
  graded bedding, 23, 46, 84
  microbial mat, 42, 112
  pyrite formation, 32, 47
  suspension settling, 82, 88, 100, 104, 106, 110
  thick lamination, 12, 23, 47, 66, 76
  tidal bedding, 66, 78
  turbidity current, 46, 82, 84, 86
pycnocline, 15, 58, 88, 90, 100
pyrite framboids
  petrography, 23, 47, 105
  SEM, 33, 45, 49, 53

random clay orientation, 27, 49-53, 59-65, 68, 70, 74, 77, 81, 85, 87, 94, 109, 111, 118
recent, 34, 64
red beds, 66
red shale, *see* red beds
regressive facies, 52, 58, 76

sabkha environment, 64
sample preparation
  petrography, 8
  SEM, 9-10
  x-radiography, 8
scanning electron microscopy
  preferred clay orientation, 25, 33, 41, 43, 45, 47, 49, 53, 61, 68, 70, 74, 77, 81, 85, 89, 94, 113, 120
  random clay orientation, 27, 49, 51, 53, 59, 61, 63, 65, 68, 70, 74, 77, 81, 85, 87, 94, 109, 111, 119
  techniques, 9-10
sedimentary environment, *see the specific environment*
sedimentary process, *see* process
sedimentation rate, 82, 93, 117
sediment gravity flow, *see* turbidity currents
SEM, *see* scanning electron microscopy
shale
  compaction, 36, 91
  definition, 2-3
  lamination, 40-41, 112
  model of forming, 94
  origin of particle orientation, 24, 94
  permeability, 117
  porosity, 117
  silt content, 95
shelf environment, 40-42, 46-52, 58-60, 72, 76-78, 100
shelf facies, 72
silt, definition, 2
slope environment, 60, 82-84
spores, 36-37, 103, 105
stratification types, 4
supratidal environment, 64
suspension settling, 82, 88, 100, 104, 106, 110

techniques, 8-9

Tertiary, 72, 82, 86, 93, 106, 110, 112, 116-117
Texas, 26, 72, 117
texture, definition, 2
thermocline, 100
thin-sections, *see* petrography
thixotropic, 93
tidal flat environment, 66-71
total organ carbon (TOC), summary, 123-128
trace fossils, *see* bioturbation

transgressive facies, 60, 76
Triassic, 62, 66
turbated sediment, 69
turbidite environment, 46, 82-86
turbidity currents, 46, 82, 84, 86, 90

Utah, 12, 26, 64, 78, 93

Van Krevelen diagram, 98-99

vitrinite reflectance, 98

West Virginia, 12, 36, 52, 93

x-radiography
  classification, 12-13
  techniques, 8